D0604318

FAR ENCOUNTER

FAR ENCOUNTER

THE NEPTUNE SYSTEM

ERIC BURGESS

COLUMBIA UNIVERSITY PRESS NEW YORK

Columbia University Press
New York Oxford

Library of Congress Cataloging-in-Publication Data
Burgess, Eric.
Far encounter : the Neptune system / Eric Burgess.
p. cm.
Includes bibliographical references and index.
ISBN 0-231-07412-3
1. Neptune (Planet)—Exploration. 2. Voyager Project. I. Title.
QB691.B87 1991
523.4′5—dc20 91-17148
CIP

∞

Casebound editions of Columbia University Press books are Smyth-sewn and printed on permanent and durable acid-free paper

Printed in the United States of America
cl 10 9 8 7 6 5 4 3 2 1

Dedication

To Phil Cleator who lead the way into space in Britain by founding the first interplanetary society there nearly 60 years ago and who, in 1936, convinced the author that spaceflight could be changed from a dream into a reality during our lifetimes.

CONTENTS

PREFACE

To me it has been a wonderful experience to have lived through what may go down in history as the greatest age of exploration in human history, an age that certainly is comparable with that of the great terrestrial voyages of discovery by Columbus, de Gama, Magellan, and Cook. How privileged I feel to have lived through this half century during which we have reached out to the far frontier of our Solar System seeking understanding of the macrocosm while other equally great explorers have delved deep within the nucleus of atoms seeking an understanding of the fundamental building blocks that form the everyday illusion of solid matter. In this preface I want to recall some of the early events leading to this exploration, particularly the early efforts to convince a doubting professional constituency that spaceflight was technically feasible and not merely an idle dream of science fiction writers.

Although Tsiolkovsky, Oberth, Esnault-Pelterie, and Goddard pioneered human thought about the technology of space exploration and the means of propulsion to achieve it, the impetus to move into space did not start until the late 1920s into the 1930s when several groups were formed with the clear objectives of pushing mankind out of the womb of Earth. In Germany Max Valier and Johanes Winkler formed the Verein fur Raumschiffahrt (1927). In the United States G. Edward Pendray and others formed the American Interplanetary Society (1930), and in Britain, Phil Cleator gathered a small group in Liverpool to form the British Interplanetary Society (1934).

Strangely, France, the home of Robert Esnault-Pelterie, did not have an active interplanetary group. When I visited Esnault-Pelterie at his laboratory at Boulogne-sur-Seine in 1937, he was chagrined that following his classic work *L'Astronautique,* no one in France had maintained an active group there. Earlier he had suggested the formation of an International Commission for Astronautics, but this had not materialized. Elsewhere, each interplanetary group individually pushed enthusiastically for space exploration, and each was strongly opposed by "experts" in many technical fields who claimed that space travel was not technically feasible.

The groups changed drastically with the advent of World War II. The German society, which by this time included a young Wernher von Braun, disappeared into military secrecy. The British Interplanetary Society notified its members that it was suspending operations for the duration of the war. The American Interplanetary Society lowered its sights and became the American Rocket Society, concentrating on rocket and jet propulsion for terrestrial applications.

In England alone the interplanetary movement continued. Following Phil Cleator's example in Liverpool, I had formed a small interplanetary group in Manchester that later became affiliated to the British Interplanetary Society just before the outbreak of hostilities. When the BIS discontinued activities I decided to keep the interplanetary movement alive and to try to hold the space enthusiasts together through quarterly meetings and a publication based on the mimeographed journal of the Manchester group with the grandiose title of *Spacewards.* As an instructor on radio and electronics with the British Royal Air Force based in rural England I was granted regular leaves. These enabled me to produce the journal and to hold meetings sufficient to keep an active membership. With support from another group, formed by Kenneth Gatland in the London area, our combined forces continued to crusade for the exploration of space. The membership of the society had grown to about 200 members by the end of the war. We had active branches in London, Manchester, and at the Royal Aircraft Establishment, Farnborough.

We also kept in close touch with the old BIS officers, Arthur Clarke, Phil Cleator, Ralph Smith, and A. M. Low. We agreed on the necessity of a national organization after the war and made plans for it. I outlined these plans to the various groups during 1945, and in December of that year official meetings of the old BIS and the Combined British Astronautical Societies wound up these groups and passed their few assets to a new Professional Society that was inaugurated under the Companies Act as a company limited by guarantee. The name of British Interplanetary Society was registered for the new company to preserve some continuity, and the Memorandum and Articles of Association were signed and a certificate of incorporation obtained in late December 1945. The society was "established for the purpose of promoting the development of interplanetary communication and travel." I became first chairman of council of the new society.

Professional societies also emerged in other countries after the war and became actively engaged in promoting the idea of spaceflight. Now the general environment had changed. While still somewhat reticent a number of engineers and scientists began to guardedly accept the possibility that what the "lunatic fringe" had been talking about for years might indeed come to pass. But I remember the comments of one reviewer, a respected scientist, of my book *Frontier to Space,* published in 1955, which reported on high altitude research by rockets and the possibility of establishing artificial satellites and sending a probe missile to orbit Mars. The final chapter, he wrote, was out of place in any scientific work since it represented an unreasonable extrapolation of current technology. In another two years the Soviet *Sputnik I* beeped its way through our heavens and the space age became a fact. So much for "expert" opinions!

The next major step forward was the organization of an international group to promote spaceflight. In 1950 Alexander Ananoff organized the First International Astronautical Congress in Paris. The handful of attendees represented a veritable who's who of astronautics with some notable absentees, mainly those who were still corralled by the military for development of long-range missiles and guided weapon systems. The Paris gathering included representatives from eight nations: Argentina, Austria, Denmark, France, Germany, Spain, Sweden, and the United Kingdom. The British delegation included Arthur C. Clarke, A. V. Cleaver, L. R. Shepherd, John Humphries and the writer. From Germany came H. H. Koelle, who was later involved in the Apollo project, and a number of officials from the Stuttgart Gesellschaft fur Weltraumforshung. Other prominent spaceflight and rocket personalities who attended included Eugen Sanger and Irene Sanger-Bredt, Rudolf Nebel, F. Schmiedl, L. Hansen, M. F. Cap, T. M. Tabanera, and A. Hjertsrand. The highlight of the Congress was a packed meeting in the Richelieu Amphitheater of the Sorbonne with stimulating speeches to an enthusiastic crowd of students and visitors.

A year later in London the International Astronautical Federation was founded at the second international congress. Representatives of the American Rocket Society also attended. The Federation became the major international body crusading for human expansion into space. It has held annual congresses every year since then at many locations worldwide.

Humankind was on its way into space. High altitude rockets sounded the upper atmosphere and beyond to reveal the complex environment surrounding our Earth. Strangely, although they had suppressed their pre-war space groups, the Soviets became the leader in establishing Earth satellites and sending probes to the Moon and the other planets; albeit sometimes not too successfully. In the ensuing years every planet in the Solar System with the exception of Pluto received an exploratory visit from at least one U.S. spacecraft.

I am awed and yet inspired by the human progress in technology during this half century but somewhat disappointed that similar progress has not been made in worldwide social and political systems. Looking back to those days of World War II just after I had received notification that the British Interplanetary Society had suspended operations for the duration, I am reminded of the evening of March 9, 1940. We had a wartime absolute blackout of all lighting in England, so conditions were ideal for astronomy. That evening a crescent moon was setting among the barrage balloons rising as ominous silhouettes against the glow of sunset to protect suburban Manchester on the outskirts of which I lived. As the sky darkened all the planets known to the ancients appeared, stretched behind the bow of the Moon like a string of beads across the evening sky: Mercury, Jupiter, Saturn, Venus, and Mars, in that order. Invisible to my unaided eye, Uranus and Neptune were also in that evening sky. But at that time I was not aware of their presence. I thought in the deep depression of those early days of the pillage of Europe that our dreams of interplanetary flight must surely remain only as dreams. Yet we recovered. Within 50 years a rocket-propelled spacecraft would travel to the frontier of our Solar System.

To me it was accordingly of great personal excitement when I first cast eyes on *Voyager* at the Jet Propulsion Laboratory. I was working on a book about the exploration of Mercury, the innermost planet of our Solar System, and was visiting the laboratory as a contractor gathering data for the book. I was invited to look at the newest project, a spacecraft to explore Jupiter and Saturn. There it stood awkwardly before me, a complex oddly shaped machine representing a merging of human innovations to carry sensors of many kinds on an unprecedented voyage of discovery.

In the years that followed I could not help admiring the creativity of the team that designed this fantastic vehicle and its instruments and the hundreds of dedicated men and women who devoted over a decade of their professional lives to this truly amazing voyage into the unknown. Surely this is an example of the triumph of the human spirit to accept a challenge and to go forward fearlessly to meet that challenge.

Perhaps with this example before us we can go forward to meet the many other challenges facing us here on Earth as we seek a global spirit, a human unity, to allow and safeguard our emergence as an interplanetary species while at the same time protecting our birthplace, the blue-green jewel of Earth, that extremely small but precious dot of consciousness in the cosmos.

Eric Burgess
Sebastopol, California

ACKNOWLEDGMENTS

The author gratefully acknowledges the assistance of the Public Information Office staff at the Jet Propulsion Laboratory; in particular, Bob Macmillan and Jurrie van der Woude. Additionally, he expresses his appreciation of the many technical presentations made by scientists and engineers connected with the projects mentioned which provided invaluable material for the writing of this book. Also, the author is indebted to Charles Kohlhase for his work in compiling the official JPL Voyager Neptune Travel Guide which provided a ready source of reference about the Voyager Project itself. Thanks are due also to Richard O. Fimmell, Pioneer Project Manager at NASA Ames Research Center and to Peter Waller of Ames Public Information Office for details of the Pioneer program's initial exploration into the outer regions of the Solar System.

FAR ENCOUNTER

1

THE DISTANT FRONTIER

This book describes the final planetary encounter of a unique spacecraft named *Voyager* which, by the wizardry of applied celestial mechanics, used swingby gravity assists to attain a far encounter at Neptune in 12 years compared with over three decades for a minimum energy transfer ellipse known as a Hohmann orbit. The opportunity for such a voyage occurs once in about 175 years, when the four outer planets, Jupiter, Saturn, Uranus, and Neptune are suitably aligned relative to each other.

The epic *Voyager* mission to the far frontier of our Solar System was conceived and the spacecraft was developed and also constructed by the Jet Propulsion Laboratory, Pasadena, California, for the National Aeronautics and Space Administration (NASA). Originally the spacecraft was named *Mariner-Jupiter-Saturn* and was intended to study the two giant planets Jupiter and Saturn since a proposed Grand Tour of the outer planets, which space planners had been working on since the 1960s, had not been approved by Congress. Critics opposing the Grand Tour contributed such inane comments as the planets would be around for another 175 years and could be explored then.

Actually the Space Science Board recommended later (in 1975) that there should be a *Mariner-Jupiter-Uranus* mission; a partial Grand Tour. This mission was scheduled for 1979 but was later canceled because of federal budget constraints. So although many people associated with the *Mariner-Jupiter-Saturn* program knew that it would be possible, if the spacecraft survived long enough, for the *Mariner-Jupiter-Saturn* spacecraft to proceed beyond Saturn to the outer planets Uranus and Neptune, this fact was not emphasized. However, it was stated that a *Mariner-Jupiter-Saturn* spacecraft might continue to Uranus if its mission at Saturn proved successful. This mission to Uranus is described in detail, together with the politics of the Grand Tour, in the author's earlier book, *Uranus and Neptune: the Distant Giants,* (New York: Columbia University Press, 1988).

Thirty times as far from the Sun as our Earth, the planet named after the god of the sea, pale-blue Neptune, is currently the outermost known planet of the Solar System. It was the final planet to be visited by the odyssey of a spacecraft aptly named *Voyager.* Planets from Mercury to Saturn were familiar heavenly objects known from ancient times as "wanderers" among the fixed stars, completing with the Sun and the Moon a mystical seven. In the seventeenth century mankind's viewpoint of the Solar System had been jolted into change by Galileo's discovery of the satellites of Jupiter and the phases of Venus. The eighteenth century produced an even more stimulating change to the human viewpoint. Unexpectedly, in 1781, the professional musician William Herschel discovered another planet, Uranus, during his systematic mapping of the stars as part of his project to determine their distances. The size of the Solar System in the mind of man had doubled because Uranus orbited the Sun more than 800 million miles beyond the orbit of Saturn, which itself was 885 million miles from the Sun. While Uranus was discovered fortuitously as the result of telescopic observations, by contrast the next important discovery resulted from mathematical analysis. Neptune was the first planet to be discovered as a result of celestial mechanics, the use of mathematical predictions based on the laws of motion and gravitation. The discovery of Neptune resulted from observations that the position of Uranus differed from what astronomers expected from the planet's orbital elements and the calculated perturbations of the then known planets. Speculations, originally by the German mathematician Friedrich Wilhelm Bessel in 1824, suggested that another planet might be causing the difference in position. About 1840 English and French mathematicians independently started to work on the involved problem of celestial mechanics to find where such a planet must be. A few years later they had each calculated the position expected for an outer planet that could account for the odd behavior of Uranus. Each sent his predictions to astronomers asking that a search should be made.

The British Astronomer Royal effectively ignored the English mathematician, John Couch Adams, and a short while later French astronomers similarly ignored the calculations of their countryman, Urbain Joseph Le Verrier. However, Le Verrier sent the results of his calculations to the Berlin Observatory in late 1846. Johann Gottfried Galle began a search there, assisted by Heinrich Louis d'Arrest. On the first night of their search, September 23, they found a planetary type object about two degrees northeast of Saturn and four degrees northeast of Delta Capricorni, a third magnitude blue-white star in the zodiacal constellation Capricornus. By the next night the faint object, which did not appear on their star charts, had moved relative to nearby stars. They had discovered Neptune. It is of interest to note speculations that Galileo had seen Neptune near the star Tau Leonis while observing Jupiter in 1613, but he recorded the distant planet as a star in his record of observations.

A search for another even more distant planet began much later when astronomers found that the presence of Neptune did not account for all the peculiarities of the position of Uranus and there were also unexplained perturbations in the motion of Neptune itself. Astronomers asked if there could be another planet beyond Neptune. Spurred by the earlier success of mathematics in showing where Neptune could be found, several mathematicians tried to predict where a trans-Neptunian planet might be located. But the main work in searching for this planet was initially undertaken by two American astronomers who were both versed in mathematical analysis, Percival Lowell at Lowell Observatory, Arizona, and William H. Pickering of Harvard College Observatory, Massachusetts. Making various assumptions they each calculated a position for a perturbing planet. Both placed it in the constellation Gemini. However, Pickering became discouraged after plates exposed at his request using a large telescope at Mt. Wilson Observatory in 1919/20 did not appear to show any planetary image. Actually, images of the trans-Neptunian planet, which would later be named Pluto, had been recorded on several plates taken at the observatory showing the area of sky in Gemini. But the star-like images had not been identified as those of a planet. Consequently Pickering abandoned the search and concentrated his work on other aspects of astronomical research.

Lowell also was unsuccessful and died in 1916 without finding his Planet-X. Ironically, as with the search by Pickering, the planet had been recorded on plates exposed at Lowell Observatory in 1915 but the faint star-like images had not been recognized as those of the elusive planet. Following Lowell's death, the search at Flagstaff Observatory ended for a while. It was resumed by Clyde W. Tombaugh soon after he joined the observatory staff in 1928. He produced an enormous number of photographic plates and checked them extremely carefully using a 'blink' comparator. He examined photographs of the star-rich regions of Gemini where Planet-X was expected to be. He searched for any faint stars that had changed position during the time between the exposure of the plates. Such objects appeared to oscillate as two photographic plates were rapidly alternated by the comparator.

Because of the many faint stars to be examined, the work was extremely tedious, but Tombaugh finally located the sought-after planet retrograding as expected in Gemini in March 1930. His success came from an indefatigable dedication and meticulous painstaking observations and inspection of photographic plates containing images of a multitude of stars. Because of his efforts Pluto joined the known planets of the Solar System. Mysteriously, though, Pluto's presence did not fully account for the perturbations in Neptune's orbit; it turned out to be much too small an object. This has led to speculations that other more distant and massive planets remain still to be discovered. Several searches have been made using a variety of methods, so far without success.

In 1846, only a few weeks after the discovery of Neptune, a British astronomer, William Lassell, discovered that Neptune has a very large satellite, Triton. Much more recently (1978) James Christy of U.S. Naval Observatory, examined photographic images of Pluto obtained with a technologically advanced 61-inch (1.55 m) astrometric reflector, located near Flagstaff, Arizona, that allowed shorter exposure times and thereby reduced image blurring caused by the terrestrial atmosphere. He saw an asymmetrical blob on an image of Pluto that he correctly interpreted as being caused by the presence of a relatively large satellite moving in an orbit very close to the planet. He had discovered Pluto's satellite, Charon. But the total mass of Pluto and Charon is only one four-hundredth that of the Earth.

Evidence appeared to mount that these four outer worlds are somehow linked harmoniously. Pluto and Charon follow an eccentric and unusually inclined orbit that carries them at times closer to the Sun than Neptune and Triton in their orbits, and farther above and below the ecliptic plane than any other planetary body (figure 1.1). Pluto revolves around the Sun in approximately 248 years, Neptune in approximately 165 years. Consequently, Pluto orbits the Sun twice while Neptune orbits three times in a stable configuration. Even though the orbits cross, a collision cannot occur because Neptune always passes Pluto when Pluto is at the part of its orbit most distant inside or outside of Neptune's orbit and far above or below the plane of Neptune's orbit. Whether these conditions will endure is questionable. Some computer simulations suggest that the orbits will change over many millions of years and become unstable. Earlier speculations that a cataclysmic event in the outer Solar System had wrenched Pluto and Triton into their unusual orbits were somewhat discounted by the discovery of Charon. Nevertheless, the four objects appear somehow to be dynamically related.

When observed through a large telescope Neptune is pale blue in color. The planet has a diameter about four times that of the Earth (figure 1.2). It orbits the Sun in 164.79 terrestrial years at a mean distance of three billion miles (4.5 billion km), about thirty times Earth's distance from the Sun. Because it is so far away and the intensity of sunlight at Neptune is almost 1000 times fainter than sunlight at Earth, the planet is too faint to be seen from Earth with the unaided eye. However, if you know precisely where to look for it with, for example, 17 × 40 binoculars, you can see Neptune as a very faint star-like object almost at the limits of visibility. Watch it for days and you will see it move relative to nearby stars. To view Neptune as a minute planetary disk you need a telescope of fairly large aperture—at least 5 inches (12.5 cm)—to provide sufficient resolving power for its 2.5 arcsecond diameter at opposition. But even the larger terrestrial-based telescopes have difficulty in revealing any details on the disk of the planet. Only broad features have been identified in Earth-based observations made since the discovery of Neptune. Nevertheless these features vary from time to time and show that the atmosphere of the planet changes dramatically compared with that of Uranus, which is generally featureless as viewed from Earth.

The equatorial diameter of Neptune was placed at 30,700 miles (49,400 km) based on observations from Earth. This classed it as slightly smaller than Uranus whose diameter is 31,760 miles (51,120 km). This accepted diameter suggested a greater density for Neptune than that of Uranus, which indicated that Neptune probably contained more heavy material than Uranus. Planetologists believed that Neptune had a two-layer internal structure; a large inner core region of rock, liquids, and ices, and an outer deep atmospheric shell of hydrogen and helium (see figure 1.2). Methane present in the planet's atmosphere contributes to the blue color of the planet by absorbing red from the reflected sunlight by which we see the planet.

An unusual attribute of Neptune is that it radiates more heat into space than would be expected if it were heated by solar radiation alone; more than Uranus radiates. Although Neptune receives 2.5 times less heat from the Sun as Uranus, the average temperature of Neptune, 60 K (−350° F) measured from Earth, is about the same as that of Uranus (59 K). Neptune radiates twice as much heat as it receives from the Sun. This implies that internal heat must be generated by the planet.

From Earth-based observations of large markings on Neptune, astronomers deduced that the planet rotates in a period of between 17 and 18 hours on an axis tilted about

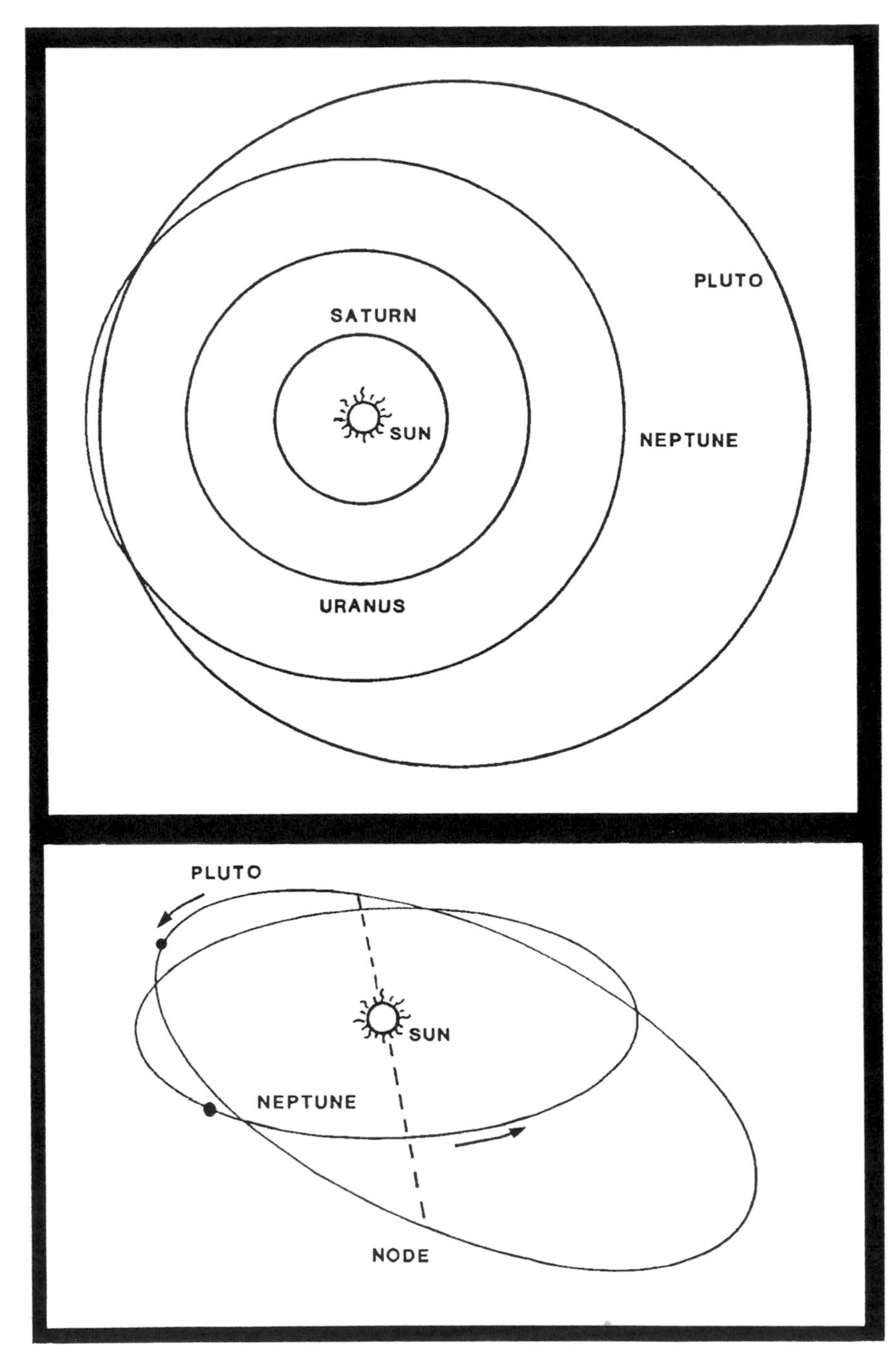

FIGURE 1.1 In a plan view of the Solar System the orbits of Neptune and Pluto appear intertwined, in fact, as shown below, the orbit of Pluto is highly inclined to that of Neptune and the two planets never approach closely to each other. Neptune revolves around the Sun three times for every two revolutions of Pluto.

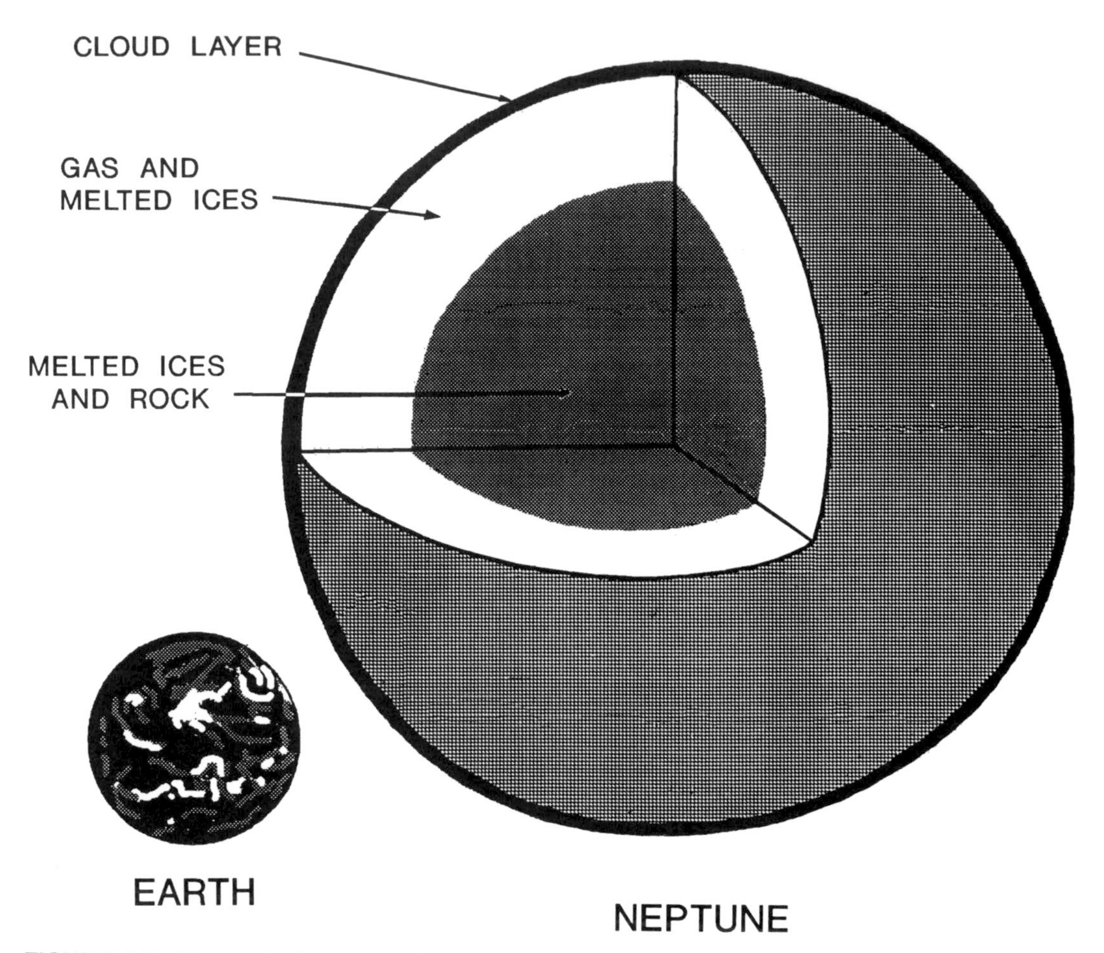

FIGURE 1.2 Neptune's diameter is four times that of the Earth; its volume almost 300 times that of Earth. Beneath the visible surface of the cloud layer a deep atmosphere merges into a shell of gas and melted ice which, in turn, becomes an inner core of melted ices and rock at high temperature and enormous pressure. It is unlikely that there is any solid core to the planet.

29 degrees to the plane of the planet's orbit. Some observations in 1988 of radio waves emitted from Neptune indicated that the planet possessed a magnetic field somewhat stronger than Earth's.

During the 1980s, when some faint stars passed behind or close to Neptune due to the motion of the planet as seen from Earth, astronomers discovered short period variations in the starlight that most probably indicated the presence of rings, or partial rings, around Neptune. While such dimming might have been caused by the presence of small satellites, the consensus was that rings were the more likely explanation. Of 110 observations of stars being occulted by Neptune, only 8 could be considered as positive evidence of starlight dimming. Five of these occultation observations are shown in figure 1.3(a). However, at least three narrow rings of 5- to 12-miles (8- to 20-km) width were generally accepted. Astronomers calculated that all the rings were in the equatorial plane of Neptune and at distances from the planet ranging from 10,500 to 26,100 miles (17,000 to 42,000 km) above the clouds.

Because the dimming of starlight was not symmetrical about the planet, and not all stellar occultations produced dimming, astronomers concluded that the rings might not be complete but only in the form of arcs (figure 1.3 [b]). Since general theories of celestial mechanics required that ring particles should spread fairly evenly and relatively quickly into a complete ring around a planet, scientists developed several theories for how partial rings or ring arcs could exist. The most popular theory was that of shepherding satellites of the type discovered at Saturn and Uranus. Such satellites could have the effect of herding ring particles and could keep them from either escaping from a ring or spreading evenly around a ring.

As *Voyager* proceeded from Uranus toward Neptune some important discoveries were made through Earth-based observations. Heidi Hammel of the Jet Propulsion Laboratory used the Infrared Telescope Facility at Mauna Kea, Hawaii, to watch the

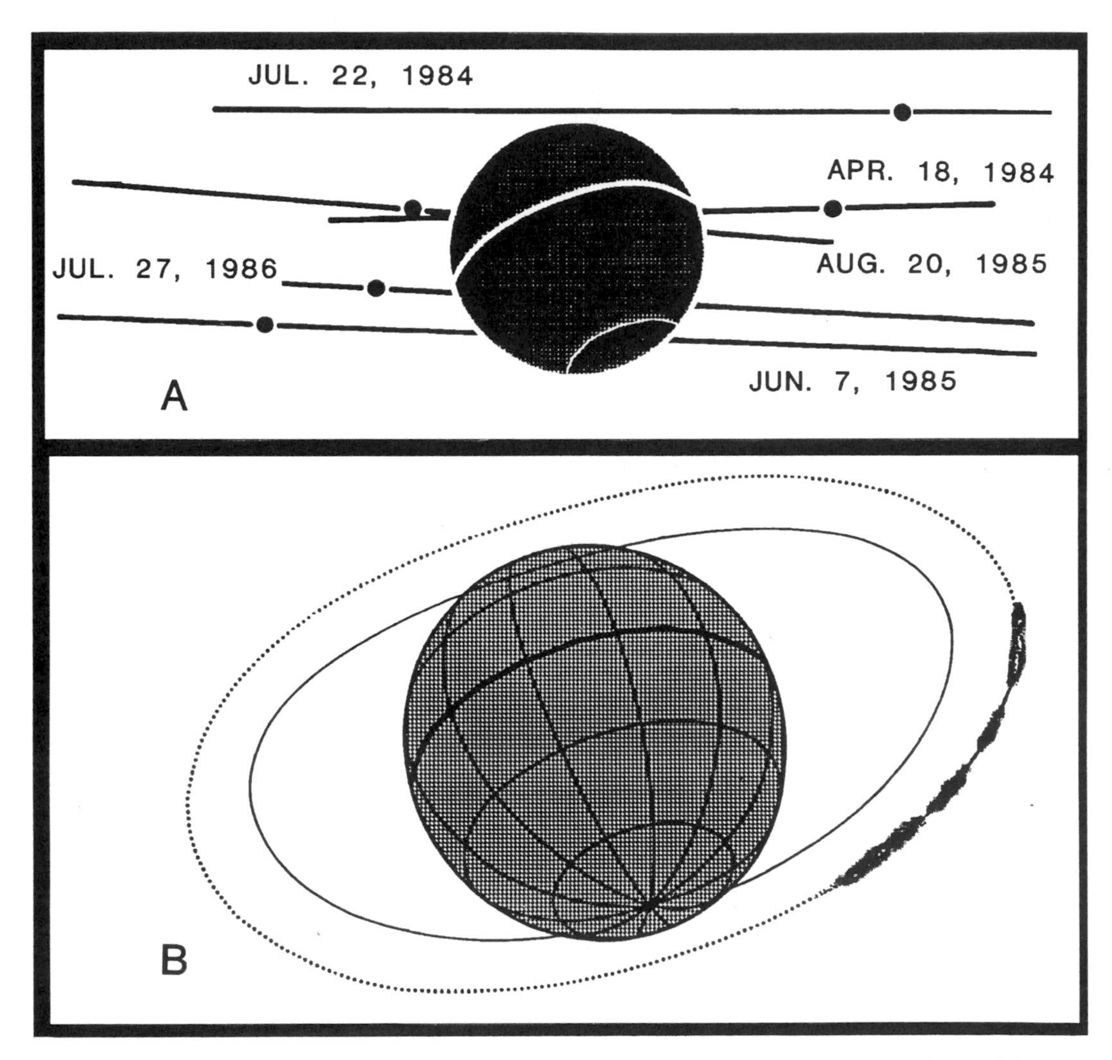

FIGURE 1.3 Several observations of stars being occulted by Neptune (A) showed the light of the star dimming for a short period before or after the star moved behind the bulk of the planet. The dates of several of these observations are shown on the drawing. These observations were interpreted as being evidence of ring arcs (B) or partial rings around the planet because the diminution of light was not symmetrical about the planet.

motions of clouds across the disc of the planet. The clouds were observed at 6190 angstroms (toward the red end of the visible spectrum) and their movement indicated that Neptune rotates at the cloud level in a period of 17 hours at 38 degrees south latitude and in 17.7 hours at 30 degrees south latitude. This indicated that wind speeds must vary at different latitudes on the planet as had been observed for Jupiter, Saturn, and Uranus. Faint hazes were also identified over the poles of Neptune.

Imke de Pater of the University of California Berkeley used the Very Large Array of radio astronomy antennas in New Mexico to measure radio emissions from Neptune. The radiation appeared to be derived from charged particles trapped in a magnetic field. This synchroton radiation confirmed that Neptune possessed a magnetic field whose strength was estimated as between 0.5 and 1 gauss, comparable to the strength of Earth's field.

Of the two known satellites of Neptune, Triton orbits at a distance of 220,160 miles (354,300 km) from Neptune in a retrograde and almost circular orbit highly inclined (157 degrees) to the planet's equatorial plane. Triton is the only large satellite traveling in a retrograde orbit around its primary. Its great distance made estimates of its exact size somewhat uncertain. The diameter had to be based on assumptions about the albedo of the surface: the percentage of incident sunlight reflected by the surface. On the assumption that the surface is ice covered, the albedo would be about 40 percent and the diameter of Triton would be about 2,200 miles (3,500 km). But depending on possible surface compositions the satellite's diameter might be as large as 2500 miles (4000 km); that is, larger than Earth's Moon but somewhat smaller than Saturn's big satellite Titan. On the other hand, its diameter could be as small as 1370 miles (2200 km), which is somewhat smaller than Titania, the largest of the Uranian satellites (figure 1.4). Evidence from Earth-based observations of the spectrum of light reflected from the satellite's surface indicated that Triton has an atmosphere containing methane. Triton appears reddish in color, which is thought to be caused by the action of the weak sunlight on chemicals in the satellite's atmosphere, somewhat similar to what is occurring on Saturn's big satellite, Titan. Also, it was speculated that if sunlight were too weak to cause such reddening, the coloration might result from cosmic ray bombardment converting methane into liquid ethane and more complex organic molecules.

There were also speculations from observations of a band in the spectrum attributed to molecular nitrogen that the satellite has nitrogen too. There were, however, no recent data to support the presence of water on Triton. In fact, it seemed that Triton was cold enough (an estimated temperature of 60 K, or −350° F) for nitrogen to exist as a liquid on the surface of the big satellite and to have nitrogen and methane frosts on solid surfaces. At that surface temperature it might be feasible for the satellite to have lakes of liquid nitrogen and methane with more complex organics floating on their surfaces. For an atmosphere of methane only the atmospheric pressure was estimated as less than 100 millibars, which is one-tenth of the terrestrial surface pressure, whereas if the atmosphere also contains nitrogen the pressure would be much lower; less than 20 millibars. Significantly, the spectrum of light reflected from Triton has changed appreciably over the last decade, presumably from seasonal effects.

Triton rotates on its axis in the same period of 5.88 days as it revolves around Neptune, so that it turns one hemisphere constantly toward Neptune as the Moon turns one hemisphere toward Earth. However, doubts had been cast on the orientation of the axis of spin of Triton. If the satellite's axis were, in fact, not perpendicular to the plane

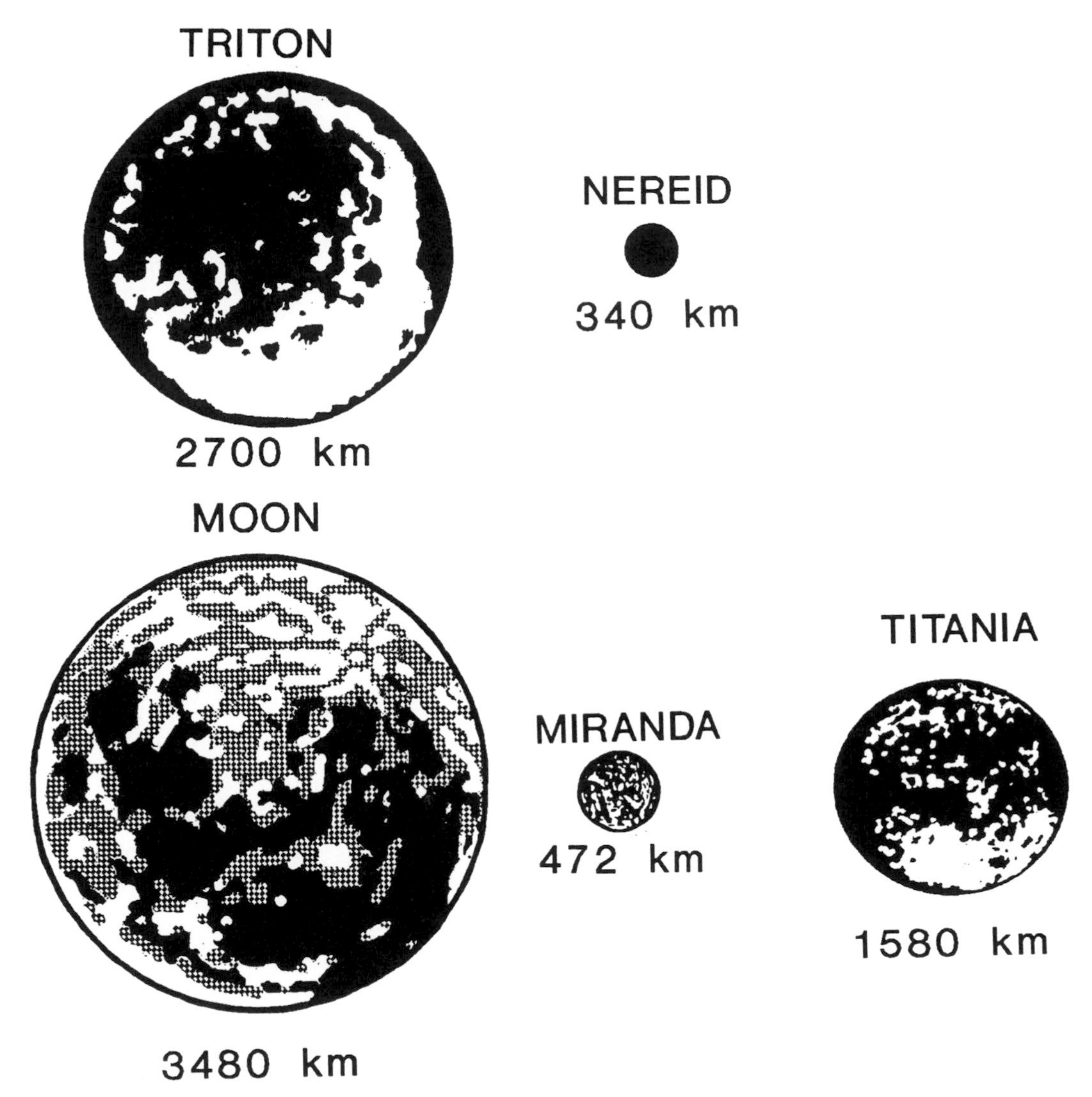

FIGURE 1.4 Before *Voyager* explored the system of Neptune, the planet was definitely known to have two satellites, large Triton and much smaller Nereid. There was also a possibility that one of the occultation observations in 1981 was that of a third satellite, but this had not been confirmed. This illustration compares the sizes of Triton and Nereid with those of Earth's Moon and Uranus's Titania and Miranda. Because the albedo of Triton was not accurately known its size might vary from larger than the Moon to as small as 2000 miles diameter. The size shown here is that determined by the *Voyager* flyby.

of the satellite's orbit, Triton might be subjected to tidal heating. This would be highest for a spin axis close to the orbital plane. In that case, with tidal heating, the surface of the satellite might have been molded by tectonic forces and volcanism.

Because of the high inclination of Triton's orbit dramatic seasonal variations were expected, since each pole would in turn face the Sun for 82 years (figure 1.5). The latitude variations of the subsolar point were calculated by L. Trafton and first published in 1984 in the journal *Icarus.* At the illuminated pole temperatures would be expected to be high enough to vaporize ices of nitrogen, methane, and argon. Also these gases would be expected to condense at the unlit pole and form a polar cap there (figure

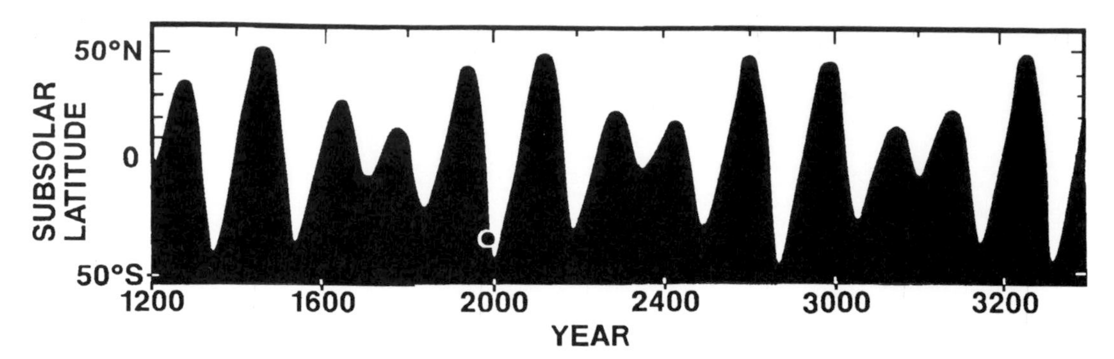

FIGURE 1.5 Triton travels in a highly inclined orbit around Neptune which precesses around the planet in a period of 640 years. Poles of Triton face the Sun in turn, leading to dramatic environmental changes on the big satellite. This illustration shows how the subsolar point (the location on the satellite where the Sun appears overhead) varies over the years, swinging from high northern to high southern latitudes. Its location at the time of the *Voyager* encounter is shown by the circle on the curve. These variations were first calculated by L. Trafton in 1984.

1.6). The density of Triton's atmosphere would accordingly be expected to vary greatly over the satellite's seasons. The tilted orbit also precesses around Neptune in a period of about 640 years, which must lead to other long term changes in how regions of Triton are exposed to solar heating. Extreme seasons are expected to occur every 640 years or so, with the next hot summer on the satellite occurring early in the twenty-first century.

Gerard Kuiper discovered a much smaller satellite of Neptune, Nereid, on photographic plates taken in 1948. Nereid is different in many ways from Triton. Its smallness and great distance from Earth made it very difficult for astronomers to estimate its exact size since even the largest telescopes show it as a mere point of light rather than a disc. Estimates based on various assumptions about its reflectivity ranged from 180 to 660 miles (290 to 1060 km) in diameter. Nereid revolves around Neptune (figure 1.7) in a normal direction but in a highly inclined (29 degrees) elliptical orbit (eccentricity of 0.75) with a period of approximately one year. Its distance from Neptune varies from 864,000 miles (1,390,000 km) at closest approach (periapsis) to 5,990,000 miles (9,640,000 km) at the most distant point of its orbit (apoapsis). No other satellites of Neptune were known at the time of *Voyager*'s approach to the distant planet, although some observations of a star occultation in 1981 led to speculations that Neptune possessed another small satellite tentatively labeled 1981N1 and orbiting close to the planet at about 47,800 miles (77,000 km).

The great distance of Neptune places severe limitations on observations even with the most modern ground-based instruments. These limitations can be removed only by spacecraft flying by or orbiting Neptune or by large telescopes in Earth orbit or on the far side of the Moon where they are free from the absorption and distortion of Earth's atmosphere. As detailed in the next chapter, *Voyager 2* was originally intended to visit Jupiter and Saturn only. It was later sent to an encounter with Uranus. Its success at that planet then provided a unique opportunity to use an existing spacecraft to explore the Neptunian system at a relatively minuscule cost of maintaining the science and spacecraft control teams for the additional years needed to travel from Uranus to the outer planet. Congressional approval for the extended mission was given, and the mission was renamed the *Voyager Neptune Interstellar Mission.* The cost of the total mission of two spacecraft for more than 12 years in space, including the encounters with all the giant planets, was now $556,000,000; which is less than a single day's cost of the defenses against an "enemy" that was fast disappearing. It was also less than one day's interest on the national debt. Per capita, the Voyager mission cost each of us in the U.S. a paltry 18 cents per year over the twelve years; less than the tax we paid on a single gallon of gasoline even before the tax increase of 1990.

Voyager had three major objectives at Neptune: to gather information about the planet, its magnetic field and magnetosphere, and Triton.

Major science questions that are extremely difficult, if not impossible, to answer from Earth, included the basic physical characteristics of Neptune, such as its size, mass, density, composition, temperature, heat balance, and rotation period. Also, basic information was required about the planet's atmosphere, cloud systems, color, temperature variations, winds, and period of rotation. Space and planetary scientists wanted to gather radio occultation data to determine the characteristics of Neptune's atmosphere to as deep a level as possible. They said a search should be made for atmospheric effects such as lightning, auroras, and radio emissions. They needed details of the magnetic field—its strength, axial inclination, offsets, and moments—for comparison with the fields of the other giant planets. They also stated that the magnetosphere should be explored to determine its extent, its interactions with the solar wind, its composition, and its interaction with any rings and small satellites, and with Triton.

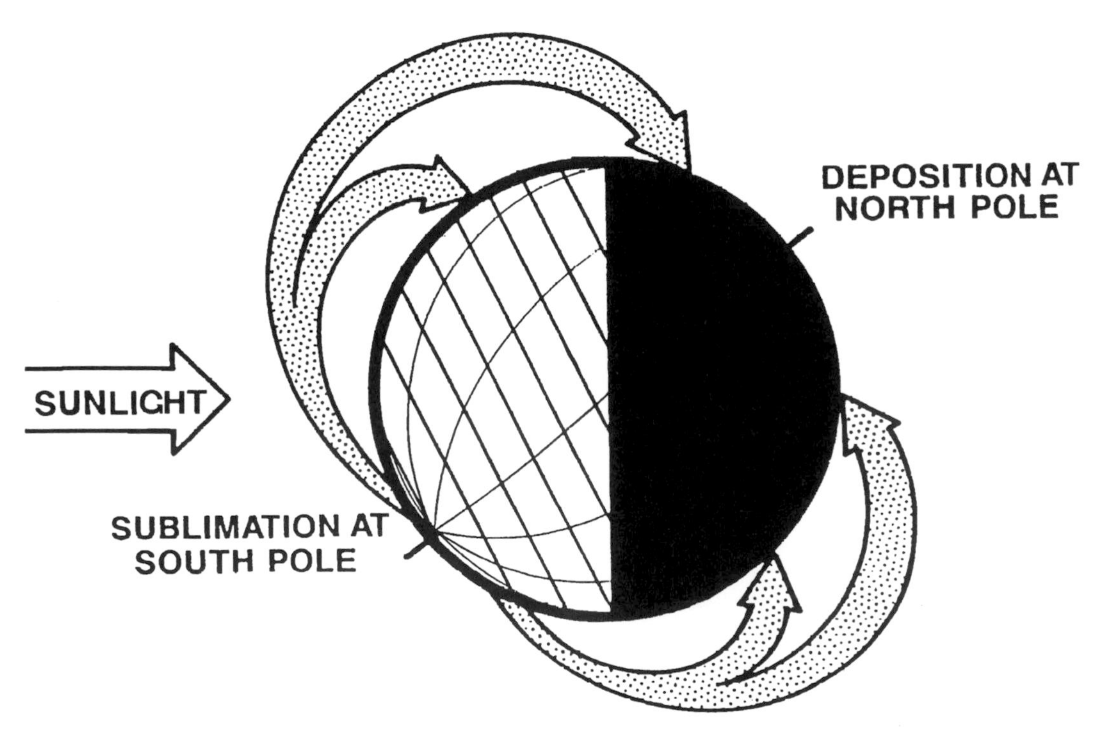

FIGURE 1.6 At the time of the encounter of *Voyager,* the southern polar region of Triton was exposed to sunlight. Heating of methane and nitrogen ices on the region causes them to sublimate. The gases migrate to the northern polar region where they condense to form an ice-covered region. This movement of ices from north to south not only changes the atmospheric pressure on a cyclical basis but also maintains a seasonal pristine ice cover and a very high albedo for parts of the satellite.

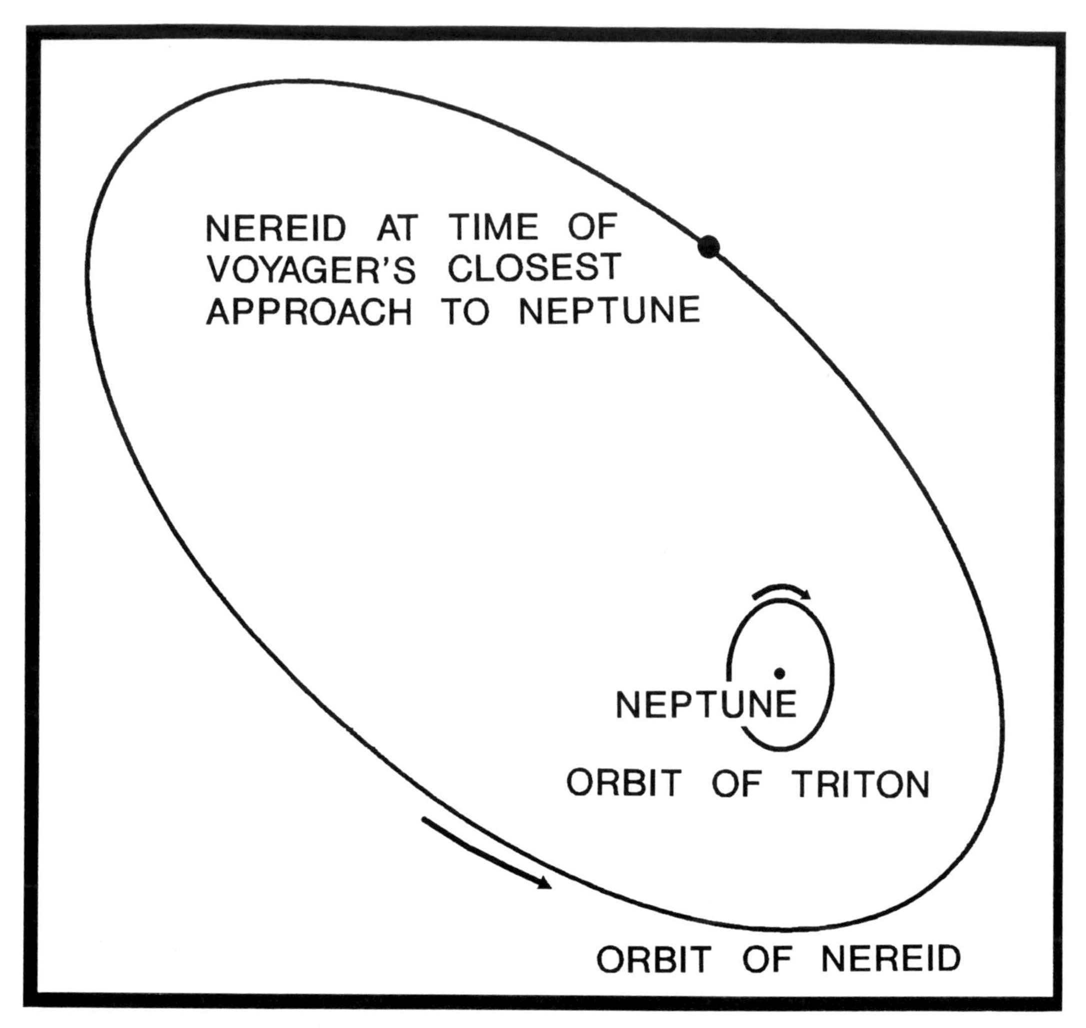

FIGURE 1.7 The two satellites of Neptune known prior to *Voyager* followed quite different orbits around the planet. Triton moves in an inclined retrograde circular orbit close to the planet with a period of just less than six days, and Nereid moves in an eccentric highly inclined orbit far from the planet with a period of almost a year. Both satellites are believed to be bodies captured by Neptune.

The intention was that *Voyager* would also be used to search for satellites and to determine the characteristics of any that were discovered. The spacecraft would search for rings and find out if there are, indeed, ring arcs and how these are configured. Shepherding satellites also would be sought, and regions of dust in the ring system needed to be looked for and charted if found.

Astronomers were also very interested in obtaining as much information as possible about the physical characteristics of Nereid and of Triton. Of great importance was the need for the spacecraft to be targeted for a close encounter with Triton to find out as much as possible about that remote and large satellite; its physical size, mass, density, and its surface features and atmosphere. An occultation of the spacecraft by Triton would provide radio astronomers with an opportunity to sound the atmosphere down to the surface of the satellite and to search for an ionosphere.

Beyond Neptune, of course, *Voyager 2* followed a trajectory out of the Solar System, one of four spacecraft outward bound to the stars. While none of these spacecraft could approach Pluto and Charon, improved ground-based equipment and a rare orbital orientation of Pluto and Charon relative to Earth have revealed much new information about this strange pair of worlds. There are many outstanding questions about these small worlds that could be answered by missions to Pluto. A possible mission suggested by the Jet Propulsion Laboratory would be launched in July 1999 and, by using swingby three times at Venus and once at Earth, could reach Pluto in 13½ years. Another mission utilizing a Jupiter slingshot, solar probe mission has been suggested for early in the next century. At the time of writing, however, there are unfortunately no approved plans for spacecraft to visit Pluto and Charon, but additional information has been gathered in recent years during the period when the pair have been inside the orbit of Neptune and hence closest to Earth and best suited for observation. Pluto reached perihelion in September 1989. In May 1990, at Pluto's closest opposition since 1742, Earth was near aphelion, and Pluto was at its brightest; about magnitude 13.5. But even so it was still nearly 800 times fainter than the faintest star visible to the unaided eye. The tiny planet could be seen as an extremely faint star-like object with the aid of telescopes of aperture greater than 8 inches (20 cm).

Charon revolves around Pluto in a period of just less than 6.39 days in an almost circular orbit at a distance of approximately 12,300 miles (19,800 km) from Pluto. The orbit is highly inclined at 122 degrees to the plane of Pluto's orbit around the Sun. Pluto's axis of rotation is inclined similarly, so that, like Uranus, it is a planet tipped on its side. While we cannot see the surface of Pluto as a disk with features from observations of which an axis of rotation could be derived, the discovery of Charon gave this important information. Gravity acting between the two closely connected bodies must bring the orbit of Charon into the plane of Pluto's equator. From the inclination of Charon's orbit astronomers thus knew the inclination of Pluto's axis of rotation.

Pluto's diurnal rotation is 6.39 days, the same as that of Charon, and the same as Charon's period of revolution around Pluto, so the two bodies continually turn to face one hemisphere to each other. Pluto and Charon went through a series of eclipses during a period of six years ending in 1991. This series of eclipses allowed astronomers to determine how much the satellite's orbit is inclined to the plane of the ecliptic. Since January 1985 Charon has been crossing in front of Pluto, transiting the planet each orbit, i.e., every 6.39 days, and behind Pluto in a similar sequence of occultations. The first observation of the eclipses was by Richard P. Binzel when on February 17, 1985, he detected a slight dimming of the light from Pluto as the edge of Charon crossed in front of part of Pluto's disk in a partial transit. Total eclipses did not occur for another two years. More was learned about Pluto during this period of eclipses than during all the 60 years since the planet's discovery.

First the uncertainty of the sizes of Pluto and Charon was resolved. Pluto has a diameter of about 1,450 miles (2,330 km), and Charon has a diameter of about 740 miles (1,190 km). Both were much smaller than had been estimated previously. But Charon is large compared with Pluto so that the pair might be described as a double planet. Only Earth has a satellite rivaling it in size. However, Earth's Moon is only about one-quarter the diameter of Earth; Charon is just over half the diameter of Pluto. Also the eclipses allowed the diameter of Charon's orbit to be refined to within a few percent; to 24,400 miles (39,280 km).

The relatively small diameter of Pluto implies that its destiny is more than had been previously thought. So it is not any longer regarded as a cometary-type ice ball but as a planet containing a significant amount of rock or of carbonaceous material, possibly of both. Astronomers interpret this density as meaning that Pluto and Charon formed as planets and not as comets way out beyond the planetary system. Whether these materials that form Pluto and Charon are homogeneously mixed with ices, or have differentiated into a core surrounded by an icy shell is not known. Charon may have a similar composition to Pluto. Only the average density of the two (about 1.9 times that of water) can be determined at present. To separate the densities astronomers would have to know the relative masses of the two bodies. This average density is comparable to that of the big satellites of Jupiter and suggests that, in contrast to the terrestrial-type planets, Pluto and Charon formed in the outer Solar System in common with the large satellites of the outer planets. The high density of Pluto has finally laid to rest the old speculative theory that a catastrophic event had pulled Pluto from its original orbit as a satellite of Neptune. Instead, Triton is now regarded as a Plutonian type body formed in the outer Solar System and captured by Neptune while Pluto and Charon escaped such a direct capture by an outer planet and entered the resonance orbit with Neptune, in which, as mentioned earlier, Neptune orbits the sun three times for every two orbital revolutions of Pluto.

What has been discovered about Pluto and Charon is thus important in seeking understanding of Voyager's discoveries at Neptune. Observations of Pluto and Charon with large telescopes during the series of eclipses also provided new information about the colors of the two bodies. When Charon was hidden behind Pluto astronomers were able to obtain a spectrum of light from Pluto uncontaminated by light from Charon. The spectrum of Charon could then be deduced by subtracting that of Pluto from the combined light from Charon and Pluto when the two were not in a position of eclipse. Charon is grayish and has a spectrum indicating that its surface is covered with water-frost, somewhat similar to that of Uranus' satellite Miranda. Methane could escape into space from the smaller Charon more easily than from Pluto.

Because Pluto is currently (1991) close to the perihelion of its orbit, solar radiation has heated some methane ice to produce an atmosphere of methane. The presence of this atmosphere is implied from strong absorption bands in the spectrum of the planet. It was confirmed in June 1988 when Pluto occulted a faint star. The light from the 12th magnitude star was not cut off abruptly but dimmed increasingly as the planet moved in front of the star. On emergence from behind Pluto after the occultation period the starlight brightened in reverse to the dimming process. The 36-inch telescope of NASA's Kuiper Airborne Observatory was used to make these observations high in Earth's atmosphere where terrestrial atmospheric effects were minimized. Pluto's atmosphere extends some 150 miles (250 km) above the surface of the planet. At the surface of Pluto the atmospheric pressure does not reach one hundred-thousandth that at Earth's surface. Much if not all of the methane in this atmosphere is expected to freeze out onto the surface as Pluto moves away from the Sun along its elliptical orbit. Astronomers expect that based on what has been discovered about Triton, Pluto's atmosphere may also contain nitrogen as the major gas species. Much of this may also freeze onto the surface as the planet moves away from the Sun.

The reddish color of Pluto implies that methane ice covering Pluto's surface has been modified. The ice would be expected to become reddened by the action of energetic particles such as cosmic rays and solar protons converting methane into colored organics. At the time of writing Pluto presents a more equatorial aspect as viewed from Earth, quite different from just after its discovery when it presented one pole toward Earth. At that period the planet was intrinsically brighter than it appears now because its apparent disk was dominated by the bright cap. This cap appears to consist of fresh methane ice that is not reddened in color. The darker equatorial zone is mottled. This mottling produces a significant cyclical change in apparent brightness as the planet rotates on its axis. When Charon obscured different parts of the disk of Pluto during the sequence of eclipses, astronomers were able to record light curves from which patches of lighter and darker areas of one hemisphere of the planet's surface could be deduced. The other hemisphere remained shrouded in mystery because the planet turns one face to Charon and turned the same face to Earth at the times of each eclipse.

Sublimation of methane ice would be expected to darken the surface of Pluto as the disappearance of light-colored frosts revealed a darker surface beneath. Indeed, over the past few decades astronomers have reported that the intrinsic brightness of Pluto has been falling as its surface darkened. The albedo of the planet will increase early next century as methane frosts are again deposited from a cooling atmosphere as Pluto's distance from the Sun increases. Pluto will again be farther than Neptune from the Sun starting in the year 2000.

The surface of Charon is different from that of Pluto and Triton. This may be because the satellite has not been able to retain an atmosphere so that there is no frozen and modified methane, and its surface is covered with water ice.

Astronomers speculate that in the outer regions of the solar nebula planetesimals could have formed bodies of relatively high density; quite different from the conditions where the icy satellites of the large planets were formed. Pluto, Charon, and Triton may represent a somewhat unique class of primordial bodies containing as much as 80 percent of rock, much of which formed from silicon and oxygen, and from which the newly formed energetic Sun had blown away most of the carbon as carbon monoxide gas. Knowing more about these very distant worlds at the far frontier is important to

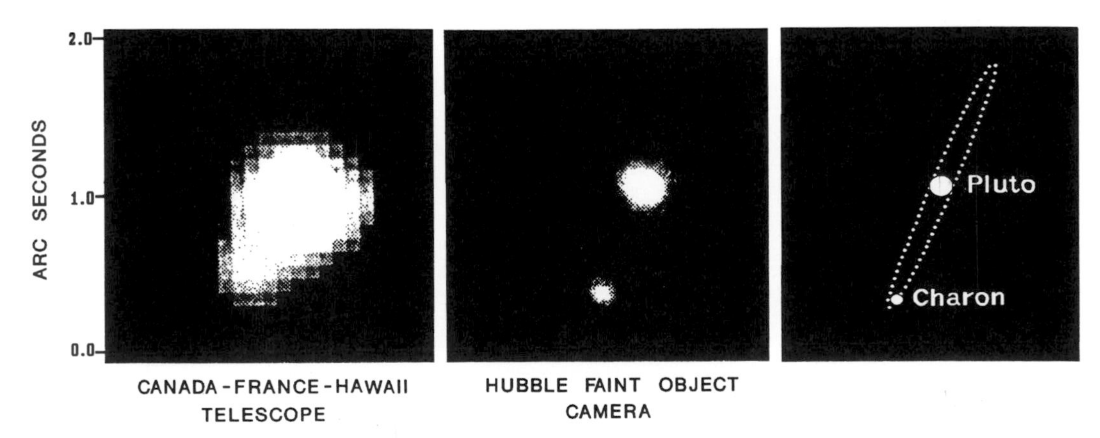

FIGURE 1.8 in 1990, the Hubble Faint Object Camera of the space telescope produced the first images showing Pluto and Charon as separate objects. By comparison the best images obtained by Earth-based telescopes, such as the Canada-France-Hawaii telescope atop Mauna Kea in Hawaii, show only an asymmetrical shape. At maximum separation Pluto and Charon are only 0.9 seconds of arc apart. (After ESA/NASA images)

our understanding of solar systems and their formation. While there are currently no spacecraft missions planned to visit Pluto in the near future, there is an opportunity in the year 2002 or 2003 to use a Mariner Mark II class spacecraft, of the type planned for an orbiting mission to Saturn, to explore Pluto. By using a swingby of Jupiter this type of spacecraft could reach Pluto after a journey of only 13 years. If this opportunity is missed there will be no further opportunities to send a spacecraft to Pluto until advanced high performance propulsion systems, such as ion drives, are available or we wait many more years for another swingby opportunity.

However, a great step forward in understanding Pluto and its satellite will come from observations with the Hubble Space Telescope which was placed in low Earth orbit on April 25, 1990. A science team led by Alan Stern of the University of Colorado plans to observe Pluto with the Wide Field and Planetary Cameras (WF/PC) and the Faint Object Camera (FOC) to decipher the puzzle of the dark and light areas on the distant planet. Images obtained with the FOC, shown at a European Space Agency press conference in Paris, November 1990, included fascinating pictures of Pluto and Charon as separate objects (figure 1.8, p. 15). The FOC was built by ESA, and of the five instruments aboard the space telescope it was the first to be verified. Its performance has been close to or better than preflight expectations. The best Earth-based telescopic photographs by comparison show the images of the two bodies merged into one asymmetrical blob. Peter Jakobsen, ESA Project Scientist, stated at the press briefing that the FOC images of Pluto and Charon were taken to coincide with their greatest separation in orbit as seen from Earth, which is close to 0.9 seconds of arc. Further observations are planned so that Charon's orbit around Pluto can be better specified. Once the orbit is accurately known, then the basic physical properties of both Pluto and Charon can be determined with greater accuracy, said Jakobsen. He explained that the telescope is not suffering from near-sightedness but rather it is like looking through foggy glasses. The images are not blurred, they are hazy, but the basic resolution is there and the haziness can be removed by computer processing.

When fully operational, the space telescope should be able to confirm if methane is deposited on the surface of Pluto as the planet moves away from the Sun and becomes cooler. Also, the telescope should be able to obtain images that will reveal details of Pluto's surface almost comparable to the details seen by the unaided eye on Earth's Moon.

Nevertheless, planning should now also be underway for spacecraft missions to the Pluto/Charon system to complete our preliminary exploration of the Solar System. Such spacecraft could be Pioneer class vehicles rather than much more expensive Voyager class vehicles. Several could be sent carrying different specialized payloads of instruments. Detailed information, obtainable only from close encounters by spacecraft is needed about the evolution of the surface, surface composition, gravity field and anomalies, magnetic fields and particles, and interactions with the solar wind. Missions should be planned to encounter Pluto before it moves far from the Sun and again when it is much more distant from the Sun to ascertain the effects of its highly elliptical orbit.

2

A UNIQUE SOLAR SYSTEM ODYSSEY

The epic space mission to the distant planets of our Solar System was originally intended to study Jupiter and Saturn only since a proposed Grand Tour of all the outer planets, which was possible only once every 175 years, was not approved by Congress. However, if the spacecraft survived the intense radiation environments of Jupiter and Saturn, it could proceed beyond Saturn to the outer planets Uranus and Neptune. Ultimately, the welcome outcome of careful design and choice of launch dates resulted in an journey of over 12 years and over 4.5 billion miles that provided a "poor man's" Grand Tour to explore the outer Solar System in this generation. This was a fantastic achievement of science and technology and human dedication to a worthwhile purpose that will most certainly be included in the history of humankind long after the petty political squabbles of this day have been forgotten.

As discussed in the first chapter, the original *Mariner-Jupiter-Saturn* mission was renamed *Voyager,* and two identical spacecraft were authorized; *Voyagers 1 and 2.* The spacecraft and the mission were ready on time with the two *Voyagers* at the Kennedy Spaceflight Center for the scheduled launch windows. *Voyager 2* was the first off the launch pad; on August 20, 1977. The second spacecraft, *Voyager 1,* was readied and it was launched soon afterward, on September 5, 1977. Each used a Titan III-E/Centaur launch vehicle. *Voyager 1* followed a faster and shorter trajectory from Earth's orbit to that of Jupiter and it arrived at Jupiter on March 5, 1979. The other spacecraft arrived there on July 9 of that year. Arrival of the spacecraft at Saturn was November 12, 1980, for *Voyager 1*, and August 25, 1981, for *Voyager 2*, to complete the nominal mission.

At launch, each Voyager spacecraft (figure 2.1) weighed 1819 pounds (825 kg) including a payload consisting of 258 pounds (117 kg) of science instruments. The spacecraft were stabilized on three axes using the celestial references of the Sun and a bright star, which was usually Canopus.

The basic structure of the spacecraft is a ten-sided aluminum framework containing ten packaging compartments. This structure is 18.5 inches (47 cm) high and 70 inches (178 cm) in diameter. A 28-inch (71-cm) diameter spherical propellant tank contained 230 pounds (104 kg) of hydrazine at launch for the 12 thrusters used to stabilize the attitude of the spacecraft and for four other thrusters used to correct its trajectory. The hydrazine was decomposed by a catalyst. The tank also contained helium at 420 psi to pressurize the hydrazine. The helium was separated from the hydrazine by a teflon-filled rubber bladder. Throughout the mission the pressure of the helium gradually decreased as the hydrazine was used.

Three radioisotope thermoelectric generators (RTGs), each contained in a protective beryllium outer case, are carried by a boom hinged on outrigger struts attached to the basic decagon structure. They provide electrical power for the spacecraft and are an essential part of spacecraft traveling to the outer Solar System where sunlight is not intense enough to power solar cells. Each RTG uses the decay of plutonium-238 dioxide to produce alpha particles the energy of which is absorbed to heat a conversion material consisting of two dissimilar materials. Electricity is thereby generated to provide 160 watts at 30 volts initially, but with a gradual decrease in power throughout the mission. The RTGs were thoroughly tested before launch to ensure that in the worst possible scenario, explosion of the launch vehicle, the plutonium dioxide would not be released into the terrestrial environment. Surplus electricity was stored during the mission in charged capacitor energy-storage banks rather than batteries. This reserve power was used for short periods to prevent the constant power voltage dipping below acceptable levels when higher than normal power drains were needed occasionally.

A major feature of each spacecraft is the 12-foot (3.66-m) diameter, high gain, parabolic antenna which is pointed almost continuously toward Earth for the receipt of commands to the spacecraft and the transmission of data back to Earth. Tape recorders on *Voyager* can store 536,000,000 bits of data. Dual frequency communication links (S-band and X-band at frequencies of 2.3 Ghz and 8.5 Ghz respectively) allowed large amounts of data to be transmitted to Earth during planetary encounters; 115,200 bits per second at Jupiter, 44,800 at Saturn, 21,600 at Uranus and 21,600 at Neptune. The 21,600 bits per second rate was maintained at the greater distance of Neptune by improvements in the ground station capabilities of the Deep Space Network. The uplink from Earth operates at S-band only whereas the downlink use both S- and X-bands.

The receiver on board the spacecraft operates continuously during the mission and can be used with either the high-gain or a low-gain antenna. *Voyager 2*'s primary receiver failed in April 1978 and the spacecraft had to operate on its backup receiver for the rest of its mission. The backup receiver suffered problems when the part of the electronics which tracks the frequency failed. This tracking loop circuit was designed to lock onto a signal from Earth to follow shifts in frequency which result from changes in velocity along the line of sight from Earth to the spacecraft because of the Doppler

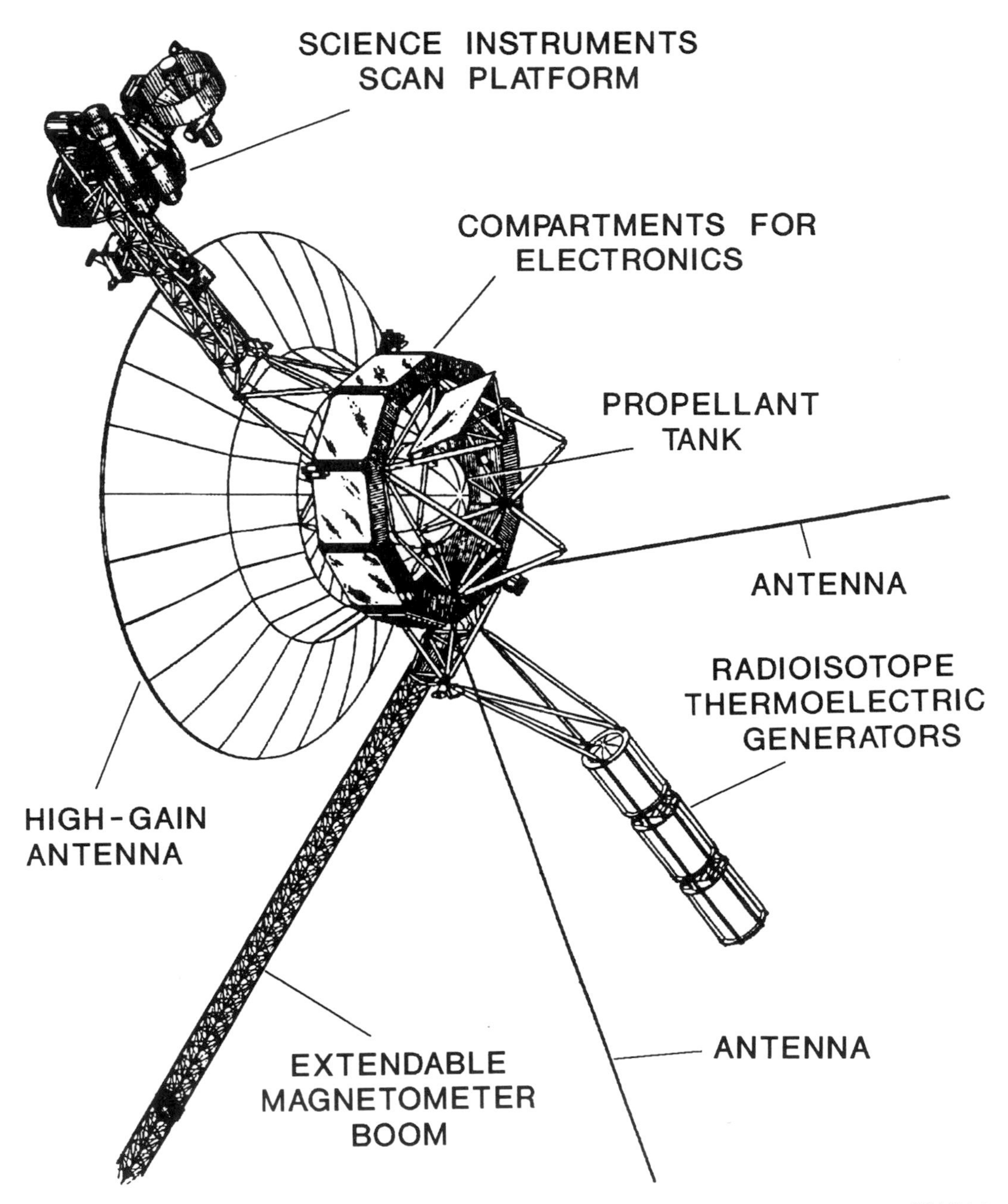

FIGURE 2.1 The *Voyager* spacecraft showing its major components and science instruments. (NASA/JPL)

effect. This had to be overcome by changing the transmitted frequency at Earth to compensate for the calculated Doppler frequency shifts on the signal received at the spacecraft. The signal at the spacecraft had to be within 96 cycles per second (hertz) of the receiver's frequency or the spacecraft would not be able to receive it. Another factor that had to be contended with was that changes in temperature within the spacecraft would change the rest frequency of the receiver. These temperature changes had to be controlled. Also, since a failure of this receiver would have made it impossible for the spacecraft to be commanded from Earth, mission controllers had to store within the spacecraft sufficient commands for it to carry out its encounter missions at Uranus and Neptune automatically. While the full mission would not have been possible using just these commands, some science data would at least have been obtained during the flybys.

The spacecraft are controlled by commands issued either in a predetermined time sequence stored in the computer on board the spacecraft or directly as received from the ground station. Such commands are transmitted digitally at the rate of 16 bits per second over the S-band link. Because of the time taken for signals to travel from Earth to the spacecraft, most commands to control operation of the spacecraft are stored within the spacecraft and issued by the computer command subsystem. Commands to control the science instruments are issued by a flight data subsystem.

Data sent from the spacecraft to Earth consists of engineering information and science information. Both types are prepared by a data storage system, a flight data subsystem, and a telemetry modulation unit for transmission to the ground stations.

Science instruments that had to be pointed were mounted on a two-axis scan platform at the end of a science boom. These included the TV cameras, spectrometers, and a photopolarimeter. The science boom also supports instruments such as the cosmic ray and low-energy charged particle detectors which are most sensitive to radiation. The boom is located opposite another, which carries the radiation-producing RTGs. A plasma science instrument is also located on the science boom.

Other instruments were mounted on the body of the spacecraft and on a third boom. Two 33-foot (10-m) "rabbit ear" antennas are for planetary radio astronomy and are also used by a plasma wave instrument. A magnetic fields experiment consists of two high-field magnetic sensors attached to the spacecraft body, and two low-field sensors mounted on a 43-foot (13-m) boom.

After *Voyager 2* encountered Saturn the motion of the scan platform in azimuth became erratic. After much diagnostic work the Mission Planning Office determined that no high speed slews could be allowed at Uranus or Neptune. Some very low speed slews would be allowed in azimuth and elevation, and a few more slightly higher speed slews would be allowed. Wherever practical, pointing of instruments would be by rolling the spacecraft.

Originally intended for a five-year mission, the *Voyager* spacecraft far exceeded their original design requirements and outperformed all expectations. During the mission all the large outer planets of the Solar System were visited and a wealth of new information was obtained about these giant worlds and their diverse families of both big and small satellites. A summary of the Voyager mission is given in table 2.1.

When the two *Voyager* spacecraft reached Jupiter in 1979, *Voyager 1* made its closest approach to the planet on March 5, and *Voyager 2* made its closest approach on July 9.

Imaging of the giant planet began in January 1979 and the first images were at a higher resolution than can be obtained by Earth-based telescopes. By early April 1979 *Voyager 1* had completed its encounter with Jupiter. In addition to much science data from its battery of instruments, the spacecraft had sent back to Earth just less than 19,000 images. *Voyager 2* continued amassing information about Jupiter from April through August, adding another 14,000 images of Jupiter and its satellites.

Many surprising new discoveries were made about the Jovian system which were not anticipated by several centuries of observation from Earth and earlier encounters by

TABLE 2.1 Summary of Voyager Mission

Planet	*Average Distance from Sun*	*Spacecraft*	*Travel Time*	*Encounter Date*	*Communication Time One Way*	*Data Rate bits/sec*	*Closest Miles*	*Approach km*
EARTH	1.00 AU	Voyager 1	Launched	5 Sep. 77				
		Voyager 2	Launched	20 Aug. 77				
JUPITER	5.20 AU	Voyager 1	18 months	5 Mar. 79	37 min.	115,200	216,796	348,890
		Voyager 2	23 months	9 Jul. 79	52 min.	115,200	448,437	721,670
SATURN	9.55 AU	Voyager 1	38 months	11 Nov. 80	84 min.	44,800	114,484	184,240
		Voyager 2	48 months	25 Aug. 81	86 min.	44,800	100,102	161,094
URANUS	19.22 AU	Voyager 2	101 months	24 Jan. 86	165 min.	21,600	66,488	107,000
NEPTUNE	30.11 AU	Voyager 2	144 months	25 Aug. 89	246 min.	21,600	18,134	29,183

two Pioneer spacecraft. Scientists obtained an enormously improved understanding of the important physical and atmospheric processes taking place on the planet and geological processes on its major satellites. New information was obtained about the near space environment surrounding the planet and about its enormous magnetosphere.

With the advent of radio astronomy in the 1950s, ground-based observers had detected radio waves from Jupiter from which they had inferred that the planet had a magnetic field and radiation belts of trapped particles similar to the terrestrial Van Allen radiation belts. The interaction of the solar wind with the magnetic field would be expected to produce a bow shock and an extensive magnetosphere. This magnetosphere's presence was confirmed, and its great extent and complexity was first revealed, by the *Pioneer 10* and *11* spacecraft which flew by the planet in 1973 and in 1974.

The Voyagers added considerably to the pioneering exploration of the Jovian system. An electric current of about 5 million amperes was confirmed flowing along the flux tube between Jupiter and Io, much greater than had been predicted before the Voyager encounter. A hot plasma of oxygen and sulfur ions was discovered near the magnetopause. Ions of these elements were also detected in a cold plasma rotating with Jupiter within 270,000 miles (430,000 km) from the planet. There were also high energy trapped particles in this same region; predominantly of oxygen, sodium, and sulfur.

A doughnut-shaped ring of plasma surrounding Jupiter at Io's orbit has plasma electron densities of more than 4,500 electrons per cubic centimeter. This torus appeared to have an oscillating plasma which gave rise to kilometric radio emissions in the range of 10 kilohertz to 1 megahertz.

In the outer magnetosphere on the day side of the planet, *Voyager*'s instruments detected plasma flows that rotated with the planet in a 10-hour period. On the night side of the planet there was evidence of a transition from closed magnetic field lines to a magnetotail, thus confirming the theoretical presence of such a magnetotail. *Voyager 2*'s observations, as it traveled from Jupiter to Saturn, indicated that the Jovian magnetotail extends some 400 million miles (650 million km) to the orbit of Saturn.

The *Voyagers* discovered that Jupiter has a ring system. The outer edge is 80,000 miles (129,000 km) from the planet's center. The ring material was not seen on the approach to Jupiter but was discovered when backlit and the spacecraft looked back at Jupiter after close encounter with the planet. The brightest outer part of the ring is about 4,000 miles (6,000 km) wide and 20 miles (30 km) thick, but much fainter ring material extends inward to the top of the planet's atmosphere. Also, diffuse ring material extends from the bright part outward to the orbit of the innermost previously known satellite Amalthea. This small satellite was found to be elliptical in shape; measuring 170 miles (270 km) by 95 miles (150 km).

Two even smaller satellites were discovered orbiting just outside the bright ring. Named Adrastea and Meris they are both about 25 miles (40 km) in diameter. A somewhat larger satellite, Thebe, was discovered orbiting between Amalthea and Io.

Of the large Jovian satellites, which were discovered by Galileo and therefore often referred to as the Galilean satellites, Europa displayed many linear features which were at first thought to be deep cracks when seen on the low resolution images returned by *Voyager 1.* The high resolution *Voyager 2* images showed no topographical features that would be expected to be associated with cracks. The lineaments appeared instead like lines drawn on a smooth surface. Their true nature is still a puzzle. However, Europa may possibly be active internally from tidal heating, and beneath a thin frozen crust of ice marked with fractures there may be oceans of water to a depth of 30 miles (50 km) or more.

The largest satellite, Ganymede, exhibits two distinct types of terrain; part is grooved and the other part is cratered. An ice-rich crust is thought to be under tension from global tectonic processes. Measurements of Ganymede's size revealed that it is the largest satellite in the Solar System, larger even than Saturn's Titan which from ground-based observations was previously thought to be the largest.

Callisto was found by the Voyagers to have an ancient, heavily cratered crust on which there are eroded concentric ring mountains surrounding very large impact basins. The largest craters on Callisto appear to have been eroded almost to the point of invisibility by the flow of ice-laden crust, and the topography of the big impact basins has almost disappeared.

Perhaps the most exciting discoveries were about Io. Nine erupting volcanoes were identified in the images of that satellite as *Voyager 1* flew past it (figure 2.2). *Voyager 2* imaged eight of these volcanoes, confirming their presence. One volcano had ceased erupting by the time of the *Voyager 2* observations. This was the first time active volcanoes had been seen in the Solar System other than on Earth. Plumes from the volcanoes rose to more than 190 miles (300 km) above the surface of Io from which scientists calculated that the material was being ejected at 2,300 miles per hour (1.05 km/sec.), 20 times the velocity at which material is ejected from Earth's most explosive volcanoes.

The volcanic activity on Io is believed to originate from heating of the interior of the satellite which resulted from the perturbations by Europa and Ganymede, which resulted in an elliptical orbit for Io, and subsequent tidal pumping by Jupiter. The action of Jupiter can produce tidal bulges on the surface of Io, with the surface rising and falling 250 feet every 42 hours; one hundred times greater than bulges on Earth's surface produced by the Moon.

A hot spot on Io associated with a volcanic feature was found to have a temperature of 290 K (60° F), which is 130 K (290° F) hotter than the surrounding surface. This spot may, indeed, be a lake of lava which may not, however, be molten, but a cooled lava surface like those found in many terrestrial lava lakes.

Io appears to be the primary source of material that pervades the magnetosphere of

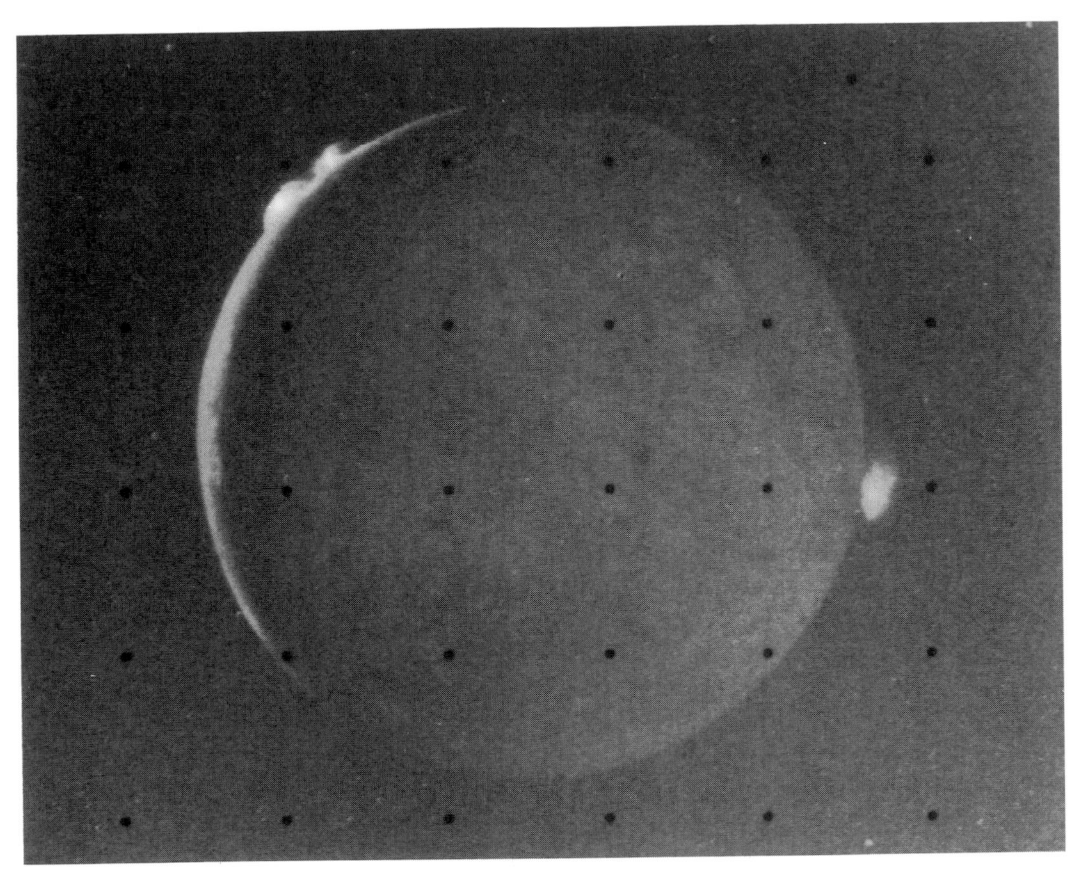

FIGURE 2.2 One of the most exciting discoveries of *Voyager*'s encounter with Jupiter was that the satellite Io has many erupting volcanoes. *Voyager 2* took this picture on July 10, 1979 from a distance of 750,000 miles (1.2 million km). The sunlit crescent of Io is to the left of the picture and the dark side of Io is illuminated by Jupitershine; by sunlight reflected from Jupiter. Three volcanic eruption plumes are visible on the limb. On the bright limb the two plumes are about 65 miles (100 km) high. The plume on the dark side limb is 115 miles (185 km) high and 200 miles (385 km) wide. The lower plume on the bright limb had increased considerably in intensity since the flyby of *Voyager 1* some four months earlier. (NASA/JPL)

Jupiter, by the volcanoes providing sulfur, oxygen, and sodium ions, and by material sputtering from the surface of the satellite as a result of high-energy particles impacting on it. Another important discovery was that conditions in the Jovian system had changed considerably since the time of the *Pioneer* flybys five years earlier. The concentrations of sulfur particles must have increased greatly since that time because *Pioneer*'s instruments had not detected them although the instruments were capable of doing so.

As for Jupiter itself, the *Voyagers* made important discoveries about the atmosphere and its features. Atmospheric features of a wide range of sizes move with uniform velocities, which suggests that atmospheric masses themselves are moving rather than the features being formed by wave motions in the atmosphere. Some of the features were seen to brighten rapidly during the period of *Voyager*'s observations, probably as the result of disturbances giving rise to upwelling and downwelling of atmospheric gases by convection. However, the winds traveling east to west extended to 60 degrees north and south latitudes. These winds are also present closer to the equator where the

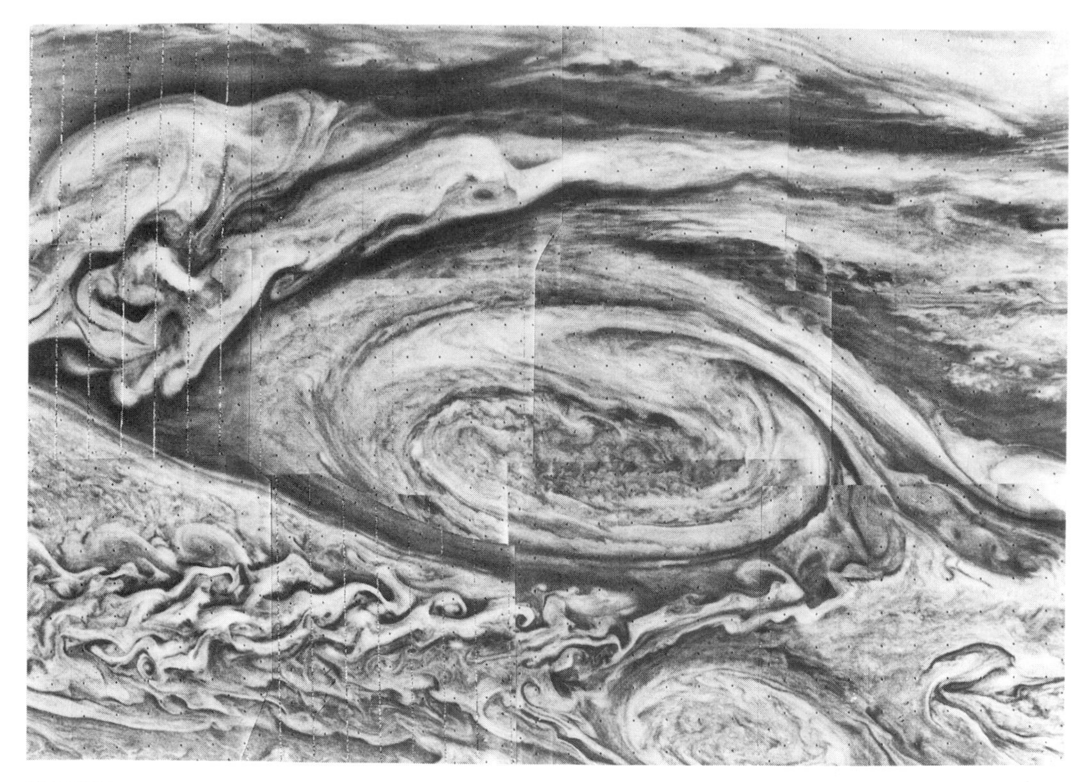

FIGURE 2.3 On Jupiter many intricate details were revealed in the images returned by the *Voyagers*. This image graphically displays the complex motions of the clouds around and within Jupiter's Great Red Spot which has been observed from Earth for several centuries. *Voyager* revealed the true nature of this mysterious marking on the giant planet as a great anticyclonic storm system. (NASA/JPL)

characteristic belts and zones are visible. The *Voyager* results indicate that convective processes are not significant compared with the winds, at least to latitudes of 60 degrees, and possibly to even higher latitudes.

The prominent Great Red Spot (figure 2.3) which has been observed from Earth for several centuries, was found to be rotating in a counterclockwise (anticyclonic) direction. The outer edge moves around the spot in a period of about five days. Near the center of the spot motions are small and random. Smaller spots at similar latitude to the big spot interact with it and with each other as they move around the planet.

The *Voyagers* found that Jupiter has auroral emissions in both ultraviolet and visible light. These emissions were not observed by the *Pioneers* and they seem to be associated with material from Io that spirals along magnetic field lines and ultimately falls into the planet's atmosphere near the poles. There are also lightning bolts at Jupiter's cloud tops.

At a pressure of five to ten millibars, the temperature of the planet's atmosphere is 160 K (−170 F), and there appears to be an inversion layer near the level where the atmospheric pressure is about 150 millibars.

The *Voyagers* also measured the temperature of the ionosphere of Jupiter and found that it reached about 1000 K (1500 F.) at maximum. This temperature was also missed by the *Pioneers*, which would indicate that there had been substantial changes to the ionosphere in the five years between the two missions.

The amount of helium in the upper atmosphere of Jupiter was found to be about 11 percent that of hydrogen by volume. This relationship is important in our attempts to understand the primordial condition of the solar nebula from which the planets formed.

In November 1980 and August 1981 the two *Voyagers* reached Saturn (figure 2.4), and were able to follow up and expand on the initial spacecraft encounter with Saturn by *Pioneer 11* in 1979. *Voyager 1* was targeted to fly by the planet along a trajectory which would allow it afterward to encounter Titan, Saturn's largest satellite and one of the very few satellites with an atmosphere. The Titan encounter curved the spacecraft's path to carry it out of the Solar System. *Voyager 2,* following the successful encounter of *Voyager 1* with Titan, could be targeted for a flyby of Saturn alone along a path that would afterward carry it into a trajectory to fly by Uranus and Neptune before the spacecraft left the Solar System.

The two highly instrumented spacecraft added considerable new knowledge about Saturn and its rings and satellites that could not be derived from Earth-based observations or from the single encounter of a much less sophisticated spacecraft, *Pioneer 11.*

Pioneer 11 made a preliminary exploration of the magnetosphere of Saturn and showed that it is a complex region affected not only by the pressure of the solar wind but also by the sweeping effects of Saturn's rings and satellites. *Pioneer* also discovered

FIGURE 2.4 *Voyager 1* looked back at Saturn four days after the spacecraft's closest approach to the planet. This unique perspective shows the shadow of the planet stretching across the magnificent ring system on which some of the mysterious spokes can be seen at the right side of the planet. (NASA/JPL)

that the magnetic field of Saturn, unlike those of other planets that possess fields, is aligned very close to the rotational axis of the planet. The magnetic axis is tipped less than one degree. Earth's field, by contrast, is inclined by 11.4 degrees. The two *Voyager* spacecraft confirmed this observation and provided more extensive and detailed data concerning the planet's magnetosphere and its interaction with the solar wind.

Several distinct regions were identified. *The Voyagers* discovered that there is a torus consisting of positively charged hydrogen and oxygen ions within approximately 250,000 miles (400,000 km) of the planet. It is believed that these ions come from water ice sputtered from the surfaces of the satellites Dione and Tethys. Associated with this torus are strong emissions from the plasma.

Toward the outer rim of the torus some ions are accelerated to high velocities giving rise to an effective plasma temperature of up to 500 million K (900 million° F).

Beyond the torus a thick sheet of plasma extends to a distance of about 600,000 miles (1 million km). The ions of this plasma are thought to arise from the ionosphere of Saturn, the atmosphere of Titan, and a region of neutral hydrogen atoms associated with Titan.

During the *Voyager* encounters with Saturn, the planet appears to have been located in the magnetotail of Jupiter; *Voyager 2* had detected this magnetotail early in 1981 as the spacecraft approached Saturn. Soon afterward, Saturn was immersed in the magnetotail, and kilometric radio emissions from the planet's magnetosphere could no longer be detected. This contrasted with the observations of the emissions by *Voyager 1* when Saturn was not in the Jovian magnetotail.

The size of a magnetosphere is controlled by the pressure of the solar wind upon it compared with the strength of the planet's magnetic field. *Voyager 2* entered Saturn's magnetosphere during a period of high solar wind pressure. At that time the bow shock, the region where the solar wind is abruptly slowed as it encounters the planet's magnetosphere, was about 710,000 miles (1.1 million km) from the planet. Several hours later the solar wind pressure fell and the magnetosphere pushed outward from the planet and remained swollen for three days at least. At that time *Voyager 2* crossed the boundary of the swollen magnetosphere as the spacecraft traveled away from the planet.

Astronomers knew that the atmosphere of Saturn consists mainly of hydrogen and helium. The relative abundances of these gases were measured by the *Voyagers.* Only 7 percent of Saturn's atmosphere is helium compared with 11 percent for Jupiter. Scientists speculated from these results that the helium in Saturn's atmosphere is sinking toward the center of the planet which, in turn, may be providing the excess heat radiated by Saturn compared with the energy it receives in radiation from the Sun.

Atmospheric markings on Saturn have long been observed from Earth as much subdued compared with those of Jupiter. This may be a result of there being more horizontal mixing in Saturn's atmosphere compared with that of Jupiter, or of less production of local color. Nevertheless, the cameras of *Voyager 2,* which were somewhat more sensitive than those of *Voyager 1,* were able to record many long-lived ovals, tilted features in east-west shear zones, and other features similar to but smaller and less distinct than those in Jupiter's atmosphere.

The winds on Saturn are moving at extremely high speeds as derived from the Voyager's images. Near the equator the winds reach 1100 miles per hour (1770 kph) blowing in an easterly direction, which is the prime direction of the winds. The dominance of eastward jet streams means that winds extend to depths of 1250 miles (2000 km) and are not confined to the cloud layer itself. The strongest winds are near the equator and the strength of the wind falls off with increasing latitude. At latitudes closer to the poles than 35 degrees the winds change direction and alternate between eastward and westward. There is also a striking north-south symmetry of the wind systems.

Measurements made of the temperature and density of the atmosphere, by analyzing radio waves that had passed through the atmosphere from the spacecraft to Earth, revealed a minimum temperature of about 82 K (−312° F) at the pressure level of 70 millibars. The temperature increased to 143 K (−182° F) at the deepest level probed by the radio waves, a point where the atmospheric pressure was about 1200 millibars; comparable to the surface pressure of Earth's atmosphere. Near the north pole the temperature at the 100 millibars pressure level was about 10 K (18° F) colder than at mid-latitudes.

There were aurora-like emissions of ultraviolet radiation at mid-latitudes and there were auroras at latitudes above 65 degrees. The high level auroral activity may be leading to formation of complex hydrocarbon molecules that are carried to the equator and produce colorations seen in the cloud bands. The auroras in sunlit mid-latitudes are puzzling because auroras are generally thought to be caused by bombardment of the atmosphere by electrons and ions, and these normally move along magnetic field lines to produce polar zone auroras.

The Voyagers also refined measurements of the rotation rate of Saturn and established it as being 10 hours 39 minutes and 24 seconds.

Some of the most surprising and significant discoveries were about the majestic rings of Saturn (figure 2.5). While the *Pioneer 11* flyby had indicated that the rings are not the simple structures that they appear to be from Earth, it remained for the Voyagers to reveal the incredible complexities of the ring system. *Voyager 2* provided the ring images of greatest resolution. An enormous amount of unsuspected detail was discovered in the classical A-, B-, and C-rings. There are many thin gaps and ringlets and very thin rings in the gaps between the classical rings. At one time it was thought that the ring gaps, such as one near the inner edge of the Cassini Division, might be caused by small satellites sweeping up ring particles. There was no trace in the Cassini Division of any satellite larger in diameter than six miles (nine km).

As light from the star Delta Scorpii passed through the rings to be recorded by the photopolarimeter of *Voyager 2,* ring structures smaller than 1000 feet (300 m) were identified from the variations in intensity of the starlight.

Scientists discovered that there are few clear gaps within the ring system. Structure seen in the B-ring, for example, appears to be caused by variations in density of the material of the rings, probably the result of traveling density waves or other waves formed by the gravitational effects of the satellites of Saturn on the ring particles. These waves propagate outward from orbits where the ring particles have a resonance with a satellite, such as a period of one-third or one-half that of the satellite. As a result the small scale structure of the rings may change frequently even though large features, such as the Cassini and Encke Divisions, have been observed from Earth for a long time.

Also *Voyager's* experiments showed that wherever gaps are present in the rings they have eccentric ringlets apparently associated with them. Variations in brightness of

FIGURE 2.5 This view shows the enormous complexity of the ring system of Saturn. It focuses on the C-ring and is made up of images taken through three filters: ultraviolet, clear, and green. More than 60 bright and dark ringlets are visible. (NASA/JPL)

these eccentric rings seem to be the result of clumping of particles, and in some parts, an almost complete absence of ring material in the orbit. Edges of rings where gaps do appear are sharp; so sharp that scientists estimate that the rings may be only about 50 feet (15 m) thick.

Two separate and discontinuous ringlets were seen in Encke's Gap in the A-ring. This gap is approximately 45,000 miles (73,000 km) above the cloud tops. One of these was found to have multiple strands.

Beyond the classical ring system, *Pioneer 11* discovered another ring, the F-ring. *Voyager* resolved this new ring into three separate strands that appear to be twisted like braids. *Voyager 2* further resolved the F-ring into five strands in a part of the ring that is not braided. The star occultation measurements revealed that the brightest of the strands in the F-ring can be further subdivided into 10 more strands. The complexities of the F-ring are believed to result from gravitational perturbations caused by small shepherding satellites, also discovered by *Voyager*. Clumps of ring material are distributed around the ring every 6000 miles (9000 km), which very nearly coincides with the relative motion of particles within the ring and the interior shepherding satellite during one revolution of the satellite.

Another surprise was the discovery of spoke-like features in the B-ring at distances between 27,000 and 35,000 miles (43,000 and 57,000 km) above the clouds. Some of the spokes are aligned radially and appear to corotate with Saturn's magnetic field in a period of 10 hours, 39.4 minutes. Other broader spokes appear to follow Keplerian orbits and rotate in the period that an orbiting body at their distance from Saturn would revolve around the planet, so that the outer end of a spoke gradually falls behind its inner end and the spoke slopes from the radius vector. Sometimes new spokes appear to be generated over the old spokes. While the spokes do not appear to form in regions near the shadow of the planet cast upon the rings, they do seem to favor a particular Saturn longitude. The spokes appeared dark as the two spacecraft approached Saturn and brighter than the surrounding ring material as the spacecraft left Saturn. This implies they consist of very fine dust particles.

Another puzzle produced by the observations from the two *Voyagers* is that the general dimensions of the rings do not remain constant at all positions around the planet. The outer edge of the bright B-ring, for example, varies in distance from Saturn by at least 90 miles (140 km), possibly even more. This probably arises because of perturbations by the satellite Mimas which has an approximately 2:1 resonance with the outer edge of the ring.

There were 11 known satellites of Saturn before the *Voyagers* reached the planet. Five had been discovered in the seventeenth century, two more during the next century, and two more in the nineteenth century. Two small satellites had been discovered by ground-based observations during the edge-on presentation of the ring system in 1966. From absorption of energetic particles a new satellite was confirmed by *Pioneer 11* in 1979. *Voyager 1* discovered more new satellites, and *Voyager 2* images added others to bring the total to 17, with possibly several more. Indeed, a very small satellite, 1981s13, was discovered nine years after the encounter of *Voyager 2*. This 12-mile (20 km) diameter satellite is in the Encke gap and was discovered in 1990, when the images were further processed by computer to search for a satellite that might be responsible for transitory variations in the nearby A-ring.

The innermost satellite in the Saturnian system was discovered by *Voyager 1*. Less than 25 miles (40 km) in diameter, it is named Atlas and it orbits near the outer edge of the A-ring. Prometheus is just inside the F-ring. It is somewhat larger (60 miles, 100 km, in diameter). Outside the F-ring the satellite Pandora, which is a few miles smaller, forms with Prometheus the pair of shepherding satellites believed to control the F-ring.

Further from the planet two larger satellites occupy almost the same orbit with a radius of 94,500 miles (151,300 km). Janus is 125 miles (200 km) in diameter, and Epimetheus is 70 miles (120 km) in diameter. Like the other small satellites they are not spherical in shape but elongated. The orbits of these two satellites differ by only about 30 miles (50 km). Consequently the satellites exchange orbits as they pass each other. One of the newly discovered satellites, Helene, shares the orbit of Dione in what is referred to as the Trojan position 60 degrees ahead of the larger satellite. Helene is about 20 miles (32 km) in diameter and is also irregularly shaped.

The larger satellite Tethys also has Trojan companions; Telesco leads and Calypso follows. These small irregular satellites have a mean diameter of about 15 miles (24 km), with Calypso being slightly smaller than Telesco. Three other small satellites were recorded on *Voyager* images but have not been confirmed by more than one image of each.

Of the larger satellites, the *Voyagers* obtained images that provided surface details impossible to obtain from Earth-based observations. The larger satellites are approximately spherical in shape. The images produced by *Voyager's* cameras revealed that Mimas, Tethys, Dione, and Rhea all have cratered surfaces. Enceladus has a surface showing great activity with five different types of surface units including some cratering, ridged plains without craters, and patterns of linear faults. The diversity of surface features is thought to be caused by tidal heating.

Mimas was shown to have a very large impact crater almost one-third the diameter of the satellite itself. Tethys, too, has a large impact crater, and the images of the satellite revealed a gigantic fracture extending almost three-quarters of the way around the satellite. Images of the satellite Hyperion show no evidence of internal activity but reveal that it is irregular in shape. The gravity of Titan periodically causes Hyperion to tumble and change its orientation. Iapetus, which was known to have hemispheres of widely differing albedo, has darker material situated on the leading hemisphere, which may imply that the material is being swept up from space onto the satellite as it revolves around Saturn. However, the floors of craters on the other hemisphere are dark and it may be that the dark leading hemisphere originates from sweeping of light material from the surface of the satellite rather than deposition of dark material on that surface.

Phoebe orbits Saturn in a direction opposite to the other satellites. Images obtained by *Voyager* showed that the satellite is roughly spherical and has a reddish colored surface. Its rotation was found not to be in synchronism with its period of revolution about Saturn, thus differing from the other large satellites. It is believed from these discoveries that Phoebe is an asteroidal body that was captured by Saturn.

The most intriguing satellite of the system is Titan, one of the Solar System's largest and believed to be the only one possessing a dense atmosphere. The chemistry of Titan's atmosphere may be similar to that of the primitive terrestrial atmosphere. *Voyager 1*'s close approach revealed that Titan is not the largest satellite in the Solar System, as it was at one time believed to be, but that it is in fact slightly smaller than Jupiter's satellite Ganymede. Radio occultation data established that the surface diameter of Titan is 3200 miles (5150 km).

The density of Titan is about twice that of water ice, which indicates that the satellite consists of nearly equal amounts of water and rock.

The surface of the satellite was hidden from *Voyager 1*'s cameras by a dense photochemical haze whose main layer is about 200 miles (300 km) above the surface. Detached layers lie above the main cloud layer with which they merge in the north polar region. *Voyager 2* discovered that there is a dark ring around the north pole and that the southern hemisphere is slightly brighter than the northern, possibly because of a seasonal effect. At the time of the Voyager flybys it was early spring in the northern hemisphere and early fall in the southern hemisphere of the satellite.

The pressure at the surface is about 1.5 times Earth's surface atmospheric pressure, and Titan's atmosphere consists mainly of nitrogen. Temperature at the surface of the big satellite appears to be about 95 K (−289° F.) so it is unlikely that rivers and lakes of methane, as had been speculated earlier, exist. However, lakes of ethane could exist on the surface in which methane could be dissolved. Methane on Titan is believed to be converted photochemically to ethane, acetylene, ethylene, and hydrogen cyanide. Hydrogen cyanide is an important molecule because it is a building block for amino acids so necessary for living organisms. However, the extremely low temperatures on Titan lead to the conclusion that molecular evolution would be unlikely to have taken place on the satellite even though the building blocks are present there.

The *Voyagers* discovered that Titan does not have an intrinsic magnetic field but it does produce a magnetic wake as it revolves around Saturn.

From the encounters with Saturn the two *Voyagers* followed very different paths. *Voyager 1,* after passing beneath Saturn to achieve an encounter with Titan, headed out of the Solar System at an angle of 35 degrees above the plane of the ecliptic; the plane of Earth's orbit around the Sun. All the planets, except Pluto, have orbits fairly close to this plane. *Voyager 2* stayed close to the ecliptic plane and headed toward an encounter with Uranus in January 1986 followed by an encounter with Neptune in August 1989.

Uranus differs markedly from most other planets of the Solar System in that its axis of rotation is tilted so much to the plane of its orbit that the planet is virtually rotating on its side. It is thought that at some time in the past the planet was hit by another planetary body that was several times larger than Earth.

Voyager approached the system of Uranus like an arrow speeding toward an archery target (figure 2.6). Before *Voyager* explored the Uranian system, scientists did not know whether or not the planet possessed an intrinsic magnetic field. Five days before the closest approach of the spacecraft, *Voyager*'s instruments detected the field's presence. But as the flight continued past the planet, scientists discovered that the magnetic field had an axis tilted some 59 degrees with respect to the planet's axis of rotation. This was by far the greatest inclination of any planetary field. The inclination causes the planet's field to wobble as the planet rotates. Also, it was found that the magnetic center of Uranus is displaced toward the surface by about one-third of the radius of the planet, much greater than for any other planet. The terrestrial displacement, by way of example, is only 0.08 Earth radii.

The magnetosphere surrounding Uranus is distorted into the shape of a windsock by the solar wind flowing past the planet; a long magnetotail extends far beyond Uranus' orbit. There are radiation belts in the magnetosphere, similar to Earth's Van Allen belts, in which high energy charged particles are concentrated.

Arising from the unusual orientation of Uranus, the planet's polar regions receive more sunlight than the equatorial regions. Were Earth aligned as Uranus is, the arctic and antarctic circles would extend almost to the equator with the whole of one hemisphere in perpetual darkness at midwinter and the whole of that same hemisphere in 24 hours of daylight at midsummer. The Tropics of Capricorn and Cancer would be close to their respective poles. Logically this was expected to cause the poles to be warmer than equatorial regions. *Voyager* discovered that this is not so. Indeed, temper-

FIGURE 2.6 (A) Artist's impression of *Voyager 2* approaching the planet Uranus with its system of thin, dark rings. The planet's polar axis points nearly in the plane of the ecliptic (plane of Earth's orbit around the Sun), and the rings appear almost fully open to Earth-based observers.

atures are fairly constant at around 64 K (−344° F) over the various latitudes, which implies that there must be a redistribution of heat within the atmosphere of the planet. Another expectation, that the atmospheric winds would differ from those of Jupiter and Saturn, was also negated by the *Voyager* encounter. Winds on Uranus have patterns much like those of Saturn. The winds flow parallel to the equator in the same direction as the rotation of Uranus. The circulation patterns are thus determined chiefly by the effects of the planet's rotation, not by the way the solar radiation is distributed on the planet.

Another discovery about Uranus was that the heat source within the planet is very much weaker than those of Saturn and Jupiter. It may be that the heat radiated from the planet is entirely derived from incident solar radiation and none from the planet's interior. Only 2 percent of the upper atmosphere of Uranus consists of methane, the gas which gives the planet its blue-green color through selectively absorbing the red component of incident sunlight. The hydrogen/helium ratio was found to be similar to the solar composition. The lower atmosphere is probably quite different in composition with as much as 50 percent water and the balance mainly methane and ammonia. The water and ammonia could not be detected by *Voyager* because they do not rise into the high atmosphere. Methane also condenses into clouds in the high atmosphere. By contrast with Jupiter, the clouds on Uranus were found to be relatively calm, optically thin, and lacking the striking features of Jupiter and Saturn clouds. Also, *Voyager* was unable to detect any lightning in the clouds of Uranus.

Many of the unanswered questions about the elusive rings of Uranus were answered during the flyby of *Voyager 2.* Nine rings had been detected from Earth in 1977 when starlight was interrupted by them during an occultation. They were also imaged from Earth by use of special electronic devices. *Voyager* imaged several additional rings and confirmed the existence of one which had been attributed to a single Earth-based observation (table 2.2).

Generally these rings are extremely narrow and they are all confined within one planetary radii of the cloud tops. They are also only a few miles thick. The rings are somewhat similar to the narrow rings discovered at Saturn. Moreover, they do not have a constant form all around the planet but are thickened or widened in parts. By contrast with the rings of Saturn, those of Uranus are dark and colorless. It has been suggested that the dark surfaces result from methane ice or carbon-rich materials modified by radiation. Alternatively the material itself may be dark like that of some asteroids.

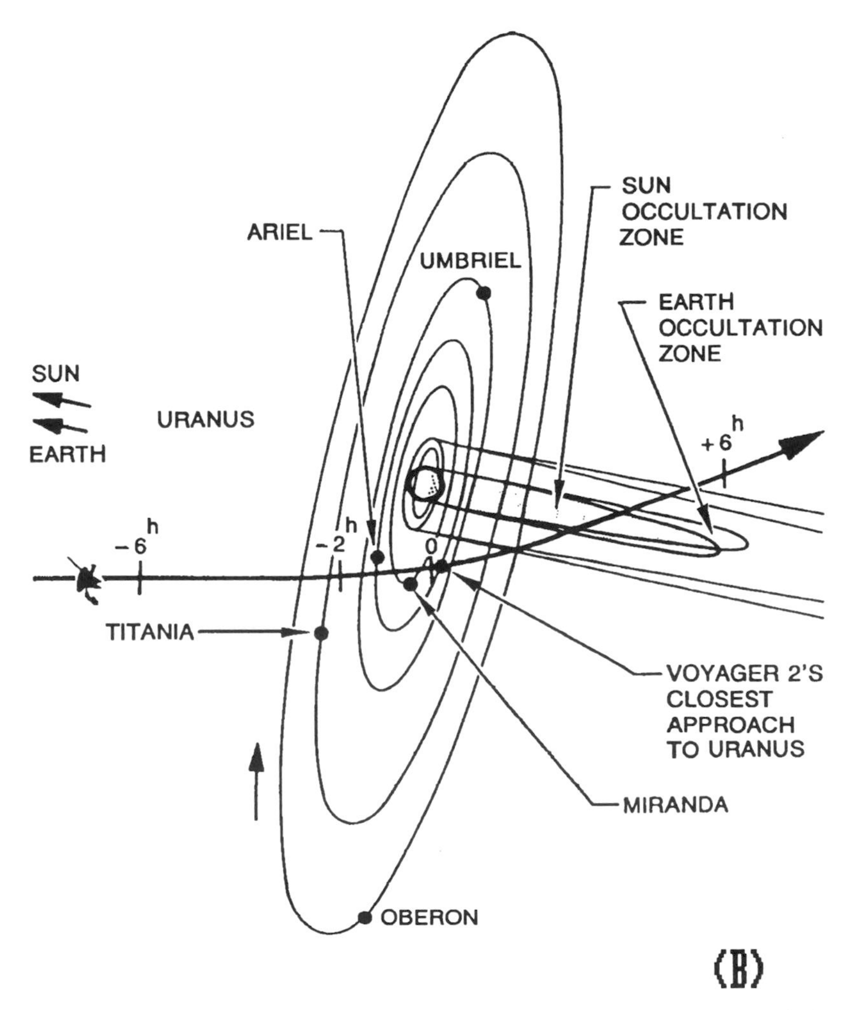

FIGURE 2.6 (B) The path of *Voyager* through the Uranian system like an arrow aimed close to a bull's eye target. (NASA/JPL).

TABLE 2.2 Rings of Uranus

Name	Max. Ring Width		Radial Distance	
	miles	*km*	*miles*	*km*
1986U2R	1,550	2,500	23,928	38,500
Ring 6	2	3	25,997	41,837
Ring 5	2	3	26,244	42,235
Ring 4	2	3	26,453	42,571
alpha ring	7.5	12	27,787	44,718
beta ring	7.5	12	28,373	45,661
eta ring	1.3	2	29,314	47,176
gamma ring	2.5	4	29,595	47,627
delta ring	5.6	9	30,012	48,299
1986U1R	1.3	2	31,084	50,024
epsilon ring	57.8	93	31,783	51,149

TABLE 2.3 Satellites of Uranus

Name	Diameter		Orbit Radius	
	miles	*km*	*miles*	*km*
Cordelia	25	40	30,915	49,752
Ophelia	31	50	33,408	53,764
Bianca	31	50	36,764	59,165
Juliet	37.2	60	38,381	61,767
Desdemona	37.2	60	38,935	62,658
Rosalind	49.8	80	39,991	64,358
Portia	49.8	80	41,072	66,097
Cressida	37.2	60	43,452	69,927
Belinda	37.2	60	46,762	75,255
Puck	105.6	170	53,442	86,004
Miranda	293	472	80,687	129,850
Ariel	720	1,158	123,967	190,950
Umbriel	728	1,172	165,295	266,010
Titania	982	1,580	271,137	436,340
Oberon	947	1,524	362,586	583,510

A surprise from the *Voyager* encounter was that the ring system of Uranus does not contain anywhere near the amounts of dust found in the ring system of Saturn. It seems that the upper atmosphere of Uranus extends a considerable distance outward from the planet and the gaseous molecules collide with ring particles to slow them until they eventually are consumed as meteors in the denser lower atmosphere. There are also shepherding satellites similar to those constraining the F-ring of Saturn. Two shepherding satellites flank Uranus' Epsilon-Ring, but no others could be found for the other rings. One explanation is that they are too dark or too small to be seen in the images returned by the spacecraft.

Voyager discovered ten additional satellites of Uranus bringing the total to fifteen (table 2.3). Of the five known from Earth-based observations two were discovered by William Herschel (the discoverer of Uranus) in 1787, two by William Lassell in 1851, and one by Gerard Kuiper in 1948. The satellites all travel in orbits approximately in the equatorial plane of the planet. Nine of the newly discovered satellites range in size from 25 to 50 miles (40 to 80 km). The tenth satellite, potato-shaped Puck, is 105 miles (170 km) in mean diameter. Eight of the satellites have orbits between those of Ophelia and Miranda, that is between the orbit of the innermost large satellite and the next large satellite outward from the planet. Two are shepherding satellites as mentioned earlier. Two satellites were large enough to be imaged as discs by the *Voyager*'s cameras. Their surfaces reflect only 7 percent of the weak incident sunlight, a very low albedo.

By comparison, the larger satellites reflect from 19 to 40 percent of incident sunlight. Even so they are darker than their counterparts at Saturn. Because the satellites are immersed in Uranus' magnetosphere for much of the time, their dark surfaces may result from the effects of energetic particles modifying organic materials on their surfaces.

Voyager 2 made important discoveries about the big satellites of Uranus. The smallest and innermost of these satellites, Miranda (figure 2.7), proved to be an extremely bizarre object when *Voyager 2* returned high resolution images of this 293-mile (472-km) diameter satellite. Areas of its surface differ greatly, possibly explained by the satellite having been impacted several times by other bodies, broken apart, and accreted again. Miranda has rolling, heavily cratered plains, a region shaped like a trapezoid with chevron-shaped light and dark markings, and oval regions of concentric canyons. The big ovals have fewer craters than the plains and are believed to have formed more recently than the other surface features. Additionally there are deep canyons with miles-high cliffs extending for great distances across the surface of the satellite.

Oberon and Titania are the largest of the satellites. Unexpectedly *Voyager* found that these satellites have greater density than Saturn's satellites. Titania's surface is redder than that of Oberon. It features complex valleys and linear faults, which suggest fracturing of an icy surface as subsurface water froze, and areas of smooth surface, which were most probably the result of ice flowing from volcanoes. In stark contrast, Oberon has a heavily cratered surface with little evidence of any internal processes having molded that surface.

The remaining two satellites, Umbriel and Ariel, are each about three-quarters the size of Oberon and Titania. Umbriel has the darkest surface. The dark coloration might be explained by assuming that the surface has been modified by collecting dark material from space resulting from the breakup of a small satellite rich in carbon. Alternatively the surface might have been dark originally. This surface is heavily cratered. A mysterious feature is that the surface does not have the young bright ray craters of the type present on the other satellites. Bright ray craters are thought to be formed when impacting bodies uncover lighter materials beneath a darker surface layer and spray the light materials radially around the point of impact. Why the surface of Umbriel does not show such young craters is a puzzle.

Ariel is the brightest of the larger satellites, and its surface is relatively free of large impact craters; but it is pitted with small craters. Since it is most unlikely that the satellite escaped heavy bombardment from space, the surface must have been smoothed by internal tectonic activity which produced volcanic flows and has left several deep chasms on the surface. Some of the larger valleys have been partially filled with smooth deposits in which there are small meandering channels. Bright areas on the surface reflect as much as 45 percent of the sunlight falling on them. These are believed to be fresh water ice spread by comets hitting the satellite.

FIGURE 2.7 A future astronaut preparing to land on the icy surface of the Uranian satellite, Miranda, might witness this awe-inspiring view of the satellite's jumbled surface and the thin lines of Uranus's rings contrasting with the bland cloud surface of the huge planet. Dominating the surface of Miranda is the great canyon with its miles' high cliffs stretching to the horizon. (NASA/JPL)

Voyager's discoveries about the large satellites of Uranus suggest that they are different in composition from the icy large satellites of Saturn. Scientists had speculated that satellites of the outer planets would decrease in density with increasing distance from the Sun because they would have accreted more volatiles, such as water and carbon dioxide, and less rocky materials. However, the density of the Uranian satellites as determined from the *Voyager* flyby implies that they contain more rocky material than do the satellites of Saturn. This rocky material would have initially had sufficient radioactive materials to heat the satellites so that they would most likely have differentiated soon after their formation. The internal heating would also account for the tectonics and volcanic flows. It is important to note that on these extremely distant worlds, where surface temperatures are extremely low, water ice behaves like rock, and when heated slightly from the interior of a satellite produces eruptive flows similar to lava flows. Volcanics on such worlds can be in the form of water and other ices rather than the fluid rocky lavas of Earth.

Voyager's launch in 1977 presented a unique opportunity to explore the system of Neptune resulting from the rare planetary configuration presented to our generation. Every 175 years the planets Earth, Jupiter, Saturn, Uranus, and Neptune attain a configuration in which their orbital positions for about three years are suitable for a spacecraft to be launched from Earth and use the gravitational slingshot of these planets to travel outward and visit each of the large planets of the Solar System. Both *Voyagers* were launched in 1977 and had the capability of traveling to all the outer planets, but scientists gave high priority to *Voyager 1* making a close approach to Saturn's largest satellite, Titan. The encounter with Titan flung Voyager 1 high above the plane of the ecliptic so that it could no longer reach Uranus, Neptune, or Pluto. Following the successful data gathering at Titan by *Voyager 1, Voyager 2* was able to use a flyby of Saturn that enabled it to travel to Uranus and from Uranus proceed on to Neptune, the outermost large planet, where Voyager's experiments led to many surprising discoveries.

3

MECHANICS OF THE FAR ENCOUNTER

At Neptune *Voyager* explored the far frontier of our planetary system with eleven scientific investigations that covered a broad range of planetary phenomena. Ten instruments were mounted at various locations throughout the spacecraft, and the communications radio doubled as a science probe to gather data about the planet's atmosphere and the atmosphere of Triton. The radio also was used to probe Neptune's ring system.

The instruments were in two categories; those that had to be pointed to achieve their purpose in gathering science data, and those that did not. Those that had to be aimed precisely were the imaging wide-angle and narrow-angle television cameras, infrared interferometer spectrometer and radiometer, photopolarimeter, ultraviolet spectrometer, and radio science. Except for the radio science subsystem, all the pointable instruments were located on the movable scan platform. The six instruments that did not need to be aimed at specific targets measured magnetic fields, energetic particles, and radio emissions. There were four magnetometers, a plasma subsystem, a low-energy charged particle detector, cosmic-ray subsystem, plasma-wave subsystem, and a planetary radio astronomy experiment. The principal investigators for all these experiments and their affiliations are shown in table 3.1. Locations of most of these experiments on the spacecraft are shown in figure 3.1.

The Imaging Science Subsystem (ISS) relied upon two vidicon cameras which

TABLE 3.1 Voyager at Neptune, Principal Investigators

Experiment	*Acronym*	*Principal Investigator*	*Affiliation*
Imaging Science	ISS	Bradford A. Smith	U. of Arizona
Infrared Interferometer Spectrometer	IRIS	Barney J. Conrath	NASA GSFC[1]
Photopolarimeter	PPS	Arthur L. Lane	JPL[2]
Radio Science	RSS	G. Len Taylor	Stanford U.
Ultraviolet Spectrometer	UVS	A. Lyle Broadfoot	U. of Arizona
Cosmic Ray	CRS	Edward C. Stone	Caltech[3]
Low-Energy Charged Particles	LECP	S. M. Krimigis	Johns Hopkins
Magnetic Fields	MAG	Norman F. Ness	U. of Delaware
Planetary Radio Astronomy	PRA	James W. Warwick	Radiophysics
Plasma	PLS	John W. Belcher	MIT[4]
Plasma Wave	PWS	Donald Gurnett	U. of Iowa

[1] Goddard Space Flight Center
[2] Jet Propulsion Laboratory
[3] California Institute of Technology
[4] Massachusetts Institute of Technology

imaged the visual appearance of the Neptunian system. In addition, the imaging system searched for ring material and for satellites of Neptune. The wide-angle camera was based on an f3.5 refracting telescope with a focal length of 7.8 inches (20 cm). Its selenium-sulfur vidicon was sensitive to light from 4,000 to 6,200 angstroms (0.4 to 0.62 microns). The narrow-angle (telephoto), f8.5 reflecting telescope camera had a focal length of 59 inches (150 cm) and covered the wavelength range from 3200 to 6200 angstroms (0.32 to 0.62 microns), extending into the near ultraviolet. It also used a selenium-sulfur vidicon to convert the visual image at the focal plane of the telescope to electrical signals. The visual range of the human eye, by comparison, is from 3800 angstroms (violet) to 6800 angstroms (deep red).

Each camera was equipped with a filter wheel that would permit different wavelengths (colors) of light to pass through while blocking light at other wavelengths. Thereby images in light of different colors could be obtained. The wide-angle camera carried a clear filter and filters of blue, green, and orange. Additionally there was a filter that passed the yellow light of Sodium-D and two filters for studying atmospheric methane. The narrow-angle camera had two clear filters, two green filters, and one filter each to pass violet, blue, orange, and ultraviolet wavelengths. Color images could be constructed at Earth by combining images taken through different filters.

The Infrared Interferometer Spectrometer and Radiometer (IRIS) was a special telescope to measure the spectrum of heat energy radiated by planets and satellites. The spectrometer was sensitive to infrared over a range from 2.5 to 50 microns and the radiometer to near ultraviolet, visual light, and infrared from 0.3 to 2 microns. Global maps of temperature distribution could be produced with this instrument. Also the instrument allowed the experimenters to study the molecular composition of atmospheres and to determine some specific atmospheric constituents.

The Photopolarimeter (PPS) was another telescope fitted with filters and analyzers of polarization; its aim was to ascertain how its targets reflect light. The properties of the surfaces can be derived from the way reflected light is polarized. The PPS provided polarization information about particles in the atmosphere and in ring systems. It was also used in a different way to investigate the ring system and atmospheres by analysis of starlight passing through the rings or the atmosphere from Sigma Sagittarii and Beta Canis Majoris. The instrument consisted of a 7.8-inch (20-cm) Cassegrain reflecting telescope equipped with eight filters and eight polarization analyzers. Unfortunately, five of the filters and four of the analyzers could no longer be used as the spacecraft approached Neptune. The instrument operated at three regions of the spectrum centered at approximately 2700 angstroms (ultraviolet), 6700 angstroms (deep orange in the visible spectrum) and 7400 angstroms (near infrared).

The Ultraviolet Spectrometer (UVS) consisted of a grating spectrometer to provide an ultraviolet spectrum from 500 to 1800 angstroms (0.05 to 0.18 microns) wavelength. It was used to gather data about the composition of the atmospheres of Neptune and Triton by determining their emission and absorption characteristics.

Figure 3.2 illustrates the frequency ranges covered by these various optical instruments over the ultraviolet, visible, and infrared regions of the spectrum.

The Radio Science Subsystem (RSS) used the radio transmissions from the spacecraft to investigate the atmospheres of Triton and Neptune and the ring system. While all the other science experiments used passive sensors which rely upon emissions, radiation, or particles coming to their detectors, the radio science was an active system. It emitted radiation and detected what happened to that radiation. When the radio beam from the spacecraft passed through the ring system and the atmospheres of Neptune and Triton, the telemetry transmissions ended and the S-band and X-band (2.3 and 8.4 GHz) carrier waves became information gatherers. Emissions from the spacecraft's transmitters were affected by passage through the planet's atmosphere or

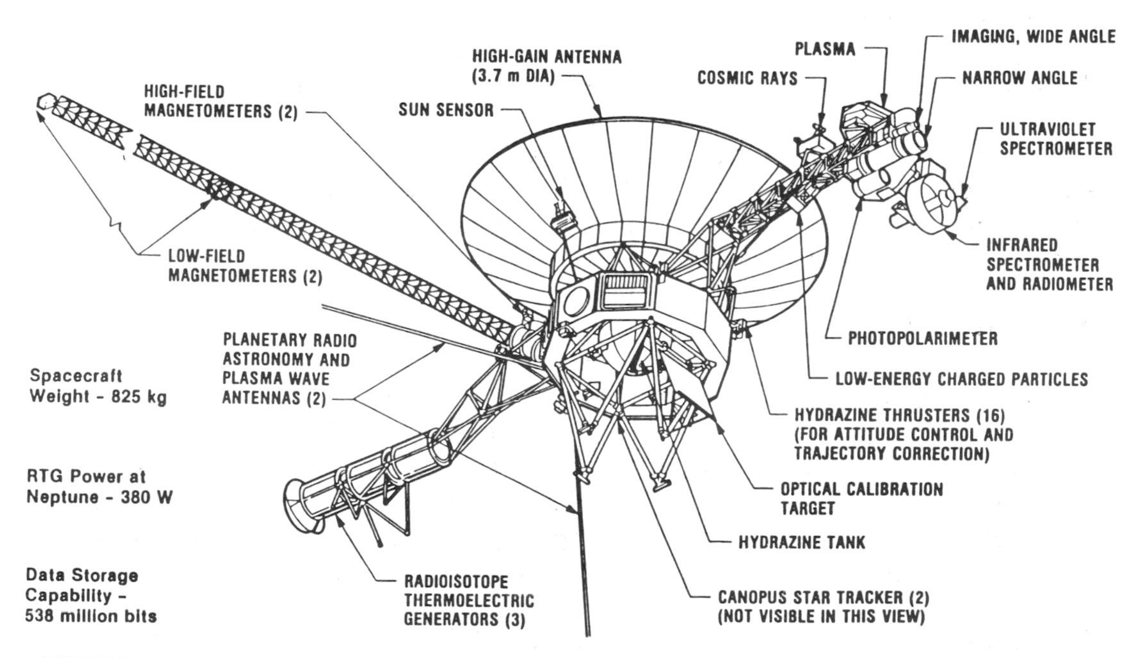

FIGURE 3.1 Locations of the science instruments on the *Voyager 2* spacecraft. (JPL)

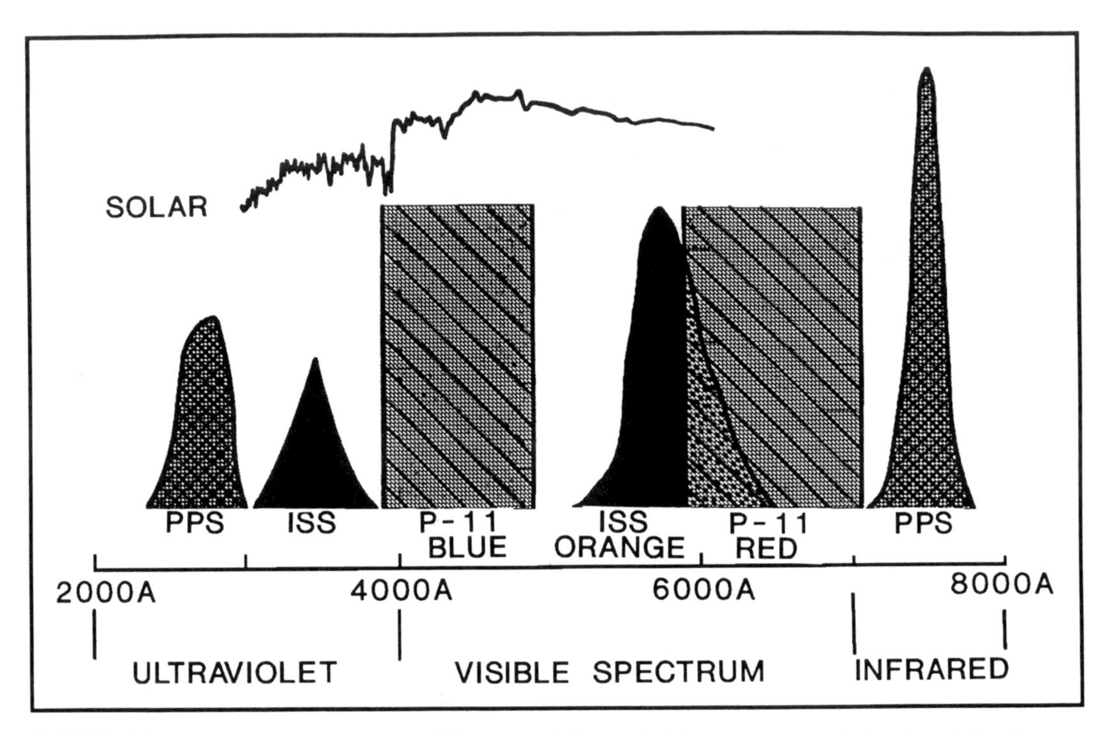

FIGURE 3.2 Frequency ranges covered by some of the optical instruments carried onboard the *Voyager* spacecraft. The top curve labeled solar shows the spectrum range of solar radiation for comparison purposes. (After JPL illustration)

the ring system and were detected by the receivers of the Deep Space Network (DSN). From changes in frequencies caused by passage through the atmosphere of a planet or a satellite, the experimenters determined the temperature, pressure, density, and composition of that atmosphere at various altitudes. When the radio signal passed through the rings, changes to its intensity and frequency allowed the experimenters to determine the number, width, shape, and thickness of the rings. Also an estimate could be made of the size of the ring particles.

Another important use of the radio science experiment was to measure the Doppler effects caused by the gravitational fields of Neptune and Triton changing the velocity of the spacecraft. This permitted experimenters to calculate the masses of the planet and the satellite more precisely.

The radio signals from the spacecraft also provided a very effective way of tracking the spacecraft through radiometrics to determine the spacecraft's location and velocity. Ranging to the spacecraft was derived from the time it took a signal from the DSN station to travel to the spacecraft and back again. The velocity along this line-of-sight was derived from the Doppler shift (change in frequency) of this signal. The angular position of the spacecraft was derived from the difference in time of arrival of its signals at widely separated DSN stations. These three measurements, made on a continuing basis through the mission, allowed navigators to determine the precise position of the *Voyager* within about 13 miles (21 km) as it approached Neptune at this far encounter over 2.8 billion miles from Earth. While this is equivalent to hitting a golf ball from Los Angeles to London and landing it within 2.5 feet of the hole, the analogy is not strictly true. Unlike the golf ball, the Voyager spacecraft was maneuvered throughout its journey—as if the ball had made close passes by Mexico City, Houston, and New York on its way to London with a few slight nudges several times during its flight.

The Cosmic Ray Subsystem (CRS) had seven independent solid-state-detector telescopes covering an energy range from 0.5 to 500 million electron volts. An electron volt is the energy acquired by an electron when it is accelerated through a potential difference of one volt. The primary purpose of the cosmic ray experiment at Neptune was to measure the energy distribution of energetic electrons trapped in the magnetosphere of the planet.

The Low-Energy Charged-Particle Detector (LECP) was intended to determine the composition, energies, and angular distributions of charged particles in interplanetary space and within Neptune's magnetosphere. It consisted of two solid-state particle detectors mounted on a rotating platform that were sensitive to charged particles with energies ranging from 15,000 to 11 million electron volts for electrons, from 20,000 to 150 million electron volts for protons and heavier ions, and from 47,000 to 200 million electron volts for atomic nuclei.

Voyager's Magnetic Fields Experiment (MAG) had four magnetometers that would allow scientists to map the magnetic field of Neptune, measure its intensity and determine its shape, and study the interaction of the field with the satellites orbiting within it. Two low-field magnetometers, which operated from 0 to 50,000 gammas (nanoteslas), were mounted on a boom holding them 43 feet (13 m) away from the spacecraft to minimize interference from fields generated by the spacecraft. The two high-field magnetometers, which can measure fields (50,000 to 2,000,000 gammas) more than 30 times stronger than that at the surface of the Earth and are less susceptible to the very weak interference, were mounted on the body of the spacecraft.

The Plasma Subsystem (PLS) was used to study the plasmas consisting of hot ionized gases that exist in interplanetary space and within the magnetospheres of the planets. The instrument consisted of two detectors that are sensitive to solar and planetary positive ions and electrons having energies between 10 and 5,950 electron volts for electrons and protons, and from 20 to 11,900 electron volts for ions. At Neptune the experiment would be used to determine the extent and shape of the magnetic field and the nature and sources of the plasma within it. Also, the experiment would check on the interaction of the solar wind with the magnetosphere of the planet.

The Plasma Wave Subsystem (PWS) was designed to measure the electric-field components of plasma waves. These waves are low frequency oscillations in the interplanetary plasma. The PWS also provided information about the interaction of plasma waves and particles that control the dynamics of Neptune's magnetosphere. Additionally, the instrument measured the density and distribution of plasma, interactions of plasma waves with energetic particles, and the interactions of satellites and rings with the magnetosphere. It could also be used to detect the impact of small particles, such as those expected in the ring plane, with the body of the spacecraft. The instrument used two extendable 33-foot (10-m) antennas which were shared with the planetary radio astronomy experiment. These were used to detect plasma waves over the frequency range from 10 hertz to 56.2 kilohertz. The experiment operated in two modes; in one it scanned frequencies over the complete band, in the other it simultaneously recorded frequencies across a band of 50 hertz to 10 kilohertz.

The Planetary Radio Astronomy (PRA) experiment used the same antennas as the plasma wave experiment to search for and establish the character of radio signals emitted by Neptune and how they relate to the satellites, the magnetic field of the planet, lightning in Neptune's atmosphere, and the general plasma environment. It operated over the frequency range of 1.2 kilohertz to 40.5 megahertz.

All Voyager flight operations and data transmissions to and from the spacecraft took place through the Mission Control and Computer Center (MCCC) at the Jet Propulsion Laboratory, Pasadena, California. Located in the Space Flight Operations Facility in Building 230 at the Laboratory, the center has four major elements; the Operations Control Subsystem, the Multi-Mission Command Subsystem, the Test and Telemetry Subsystem, and the Information Processing Center.

Commands for the spacecraft and its systems and science experiments were processed and sent from the MCCC via the Ground Communications Facility, which is operated by NASA's Goddard Spaceflight Center, Maryland, to the stations of the Deep Space Network (DSN). From the huge antennas of these stations the commands to Voyager were beamed across the billions of miles of space.

Data from the spacecraft were received by these same antennas of the Deep Space Network and from there were passed by NASA's Ground Communications Facility to the computer center at the Jet Propulsion Laboratory. The data were processed at the Space Flight Operations Facility by one or more of several computers needed to organize it into appropriate formats for display and for distribution to and analysis by engineers and scientists associated with the program. All data from the spacecraft's science experiments were collected into Experimental Data Records which became the final data product forwarded to investigators. There was also a supplementary record that gave the experimenters a best estimate of the conditions under which the scientific observations were made by the spacecraft's instruments.

The Deep Space Network's large antennas are located around the world at widely separated longitudes so that a spacecraft can always be kept in view from one or another of the stations as Earth rotates on its axis. There are three complexes of DSN; one is at Goldstone, near Barstow, in the Mojave Desert of California, a second is at Robledo, near Madrid, Spain, while the third is at the Tidbinbilla Nature Preserve near Canberra, Australia. Each DSN station supported the Voyager mission with three big antennas. The giant of the system at each station is a 230-foot (70-m) diameter parabolic antenna which can both transmit and receive. At each station there are also two smaller antennas of 112-foot (34-m) diameter. One of these, referred to as the standard antenna, is capable of transmitting and receiving, while the other, a high efficiency system operating at X-band, only receives signals. The 230-foot and the standard 112-foot antennas can receive signals at S- and X-band (2.3 GHz and 8.5 GHz) simultaneously. Each of the 230-foot antennas in the DSN has a 400-kilowatt transmitter, whereas each standard 112-foot antenna transmits at a power of 20 kilowatts.

The transmitter on the *Voyager* spacecraft beamed its radio signals earthward across the billions of miles of space toward the receivers attached to the big DSN antennas at a mere 20 watts of power (half that of a typical high-intensity desk lamp). Imagine trying to see the beam from a small flashlight at a distance of 3 billion miles!

The radio beam spreads on its long journey, so after the signals had traveled 3 billion miles, the big antennas of the DSN had to handle an electromagnetic whisper whose received power was 20 billion times less than the power used by a modern digital watch. The received signal had a minuscule power of only 10^{-16} watts, and was processed by maser amplifiers which amplified the signal with minimum introduction of electronic noise into it.

So that they could accomplish this amazing feat of communication, the antennas were upgraded following the encounter of *Voyager* with Uranus. To provide a greater receiving area and a more accurate surface to focus the faint signals, the steel structures of the antennas were stripped and replaced. The diameter of each dish was increased from 210 to 230 feet (64 to 70 m) to receive the incredibly weak signal arriving from the spacecraft at Neptune. The signals were processed by ultra-low-noise maser amplifiers.

For the encounter with Neptune the 230-foot and 112-foot antennas were linked together so that their received signals could be combined in a process known as arraying. At Canberra the antennas were also linked with a 210-foot (64-m) antenna at Parkes Radio Astronomy Observatory that was fitted with a maser on loan from the European Space Agency. The addition of this antenna not only increased the capability of receiving faint signals at higher bit rates but also minimized terrestrial interference. The Parkes and the DSN Canberra antennas are about 200 miles (320 km) apart and less subject as a pair to local interfering conditions such as thunderstorms. Heavy clouds, rain, or snow, and even high humidity absorb the 8.5 MHz signals and cause loss of data.

The Canberra station of the DSN played a major role during the encounter of Voyager with Neptune, as it had during *Voyager*'s Uranian encounter, because at the time of both encounters Neptune and Uranus were located in Sagittarius, a zodical constellation below the celestial equator. The celestial equator is the projection of Earth's equator on the star sphere. The Sun appears in this constellation during midwinter in our Northern Hemisphere, and is always very low in the winter sky. Sagittarius rises high in the terrestrial sky only as seen from Earth's Southern Hemisphere. At Canberra the Sagittarius region is above the horizon sufficiently long for 12 hours of good communication with the spacecraft each day. At Goldstone, by comparison, the available communication period each day is less than half that at Canberra.

To safeguard the science data from the mission two other antenna systems, in addition to the Parkes Radio Telescope, supplemented those of the Deep Space Network. In New Mexico the twenty-seven, 82-foot (25-m) antennas of the Very Large Array (VLA) of the National Radio Astronomy Observatory were arrayed with the Goldstone antennas to provide more capability to receive the extraordinarily weak radio signals from the spacecraft.

The VLA is a radio astronomy antenna system operated by Associated Universities, Inc. for the National Science Foundation. The partnership with the Voyager project started in 1982 when mission planners recognized that if the *Voyager 2* spacecraft survived long enough to fly by Neptune, some method of increasing the signal-receiving capability would be needed for reasonable data rates to be achieved. The VLA is a Y-shaped arrangement of big antennas on a plain about 50 miles (80 km) west of Socorro. The system is normally used to detect and map fine details in distant radio sources such as galaxies. Each 230-ton antenna can be moved along railroad tracks to selected locations on the Y-pattern to form clustered or spaced out configurations for various uses of the VLA. The antennas were placed in a cluster pattern for arraying with the DSN station.

Some improvements to the VLA system were necessary for its operation with *Voyager* at Neptune. Over a period of four years, highly efficient receivers, based on high electron mobility transistors were added to each antenna so that it could receive the X-band signals from the spacecraft. Also, auxiliary power generating systems were added to the VLA system to ensure that uncertain utility power transients and outages would not jeopardize the receipt of images from *Voyager* during the critical time of encounter at Neptune.

In Japan the Institute for Space and Astronautical Science (ISAS) has a 210-foot (64-m) radio telescope at the Usada Radio Observatory. For the Neptune encounter this antenna system used a hydrogen maser on loan from NASA, and was connected via communications to the Canberra DSN station to collect S-band radio science data for the very important periods of the precisely timed radio occultation experiments.

Commands sped to the spacecraft at a rate of 16 bits per second. They took an unprecedented 4 hours and 6 minutes to travel from Earth to *Voyager* at Neptune. A round trip time between Earth and *Voyager* was thus over 8 hours, and with such a long communication delay engineers could not respond quickly if the spacecraft had developed problems. Accordingly, a master computer carried by *Voyager,* the Computer Command Subsystem, was programmed with a set of stored responses to anticipated problems. By this means the spacecraft would act of its own accord to protect itself from situations that could interrupt operations of the spacecraft or break communications with Earth. Stored within the on-board computer was a backup mission load containing basic commands that would have allowed *Voyager* to conduct rudimentary science experiments at Neptune if communications with the spacecraft had failed. This backup was necessary primarily because the communications receiver of *Voyager 2* had been giving trouble for several years and was being nursed through the mission. Also, the backup guarded against interruption of transmission from the Earth end of the communication link.

At times during the mission the spacecraft had to be given what are termed real-time commands. These are commands sent for action immediately on their receipt at the spacecraft. They were used when the spacecraft or its instruments had to be commanded other than by commands already stored within the spacecraft's computer memory.

There were several different phases established for the encounter as summarized in table 3.2. Each phase had specific science objectives best suited for the location of the spacecraft with respect to the Neptune system. Most activity took place, of course, during the relatively short period of close encounter with Neptune and Triton.

Voyager officially began its encounter with Neptune on June 5, 1989, which was 81 days before the closest approach to the planet. The spacecraft was then 73 million miles (117 million km) from the planet. Just over two months were spent in an observatory phase during which the spacecraft underwent many checkouts of its various systems and instruments. When each had been checked and calibrated as needed, the science instruments began to amass important new data. Scans across the Neptune system were made with the ultraviolet spectrometer searching for evidence of neutral hydrogen and ions. The imaging system's cameras provided a sequence of images of the planet containing valuable information about atmospheric motions. Also a search began for rings and small satellites. Toward the end of the observatory phase the trajectory received slight corrections. On August 6, 1989 the far encounter phase began and lasted until August 24. At the beginning of this phase the planet loomed large in the

TABLE 3.2 The Phases of Voyager's Encounter with Neptune

Event	*Date*	*Earth Received Time (PDT)*
1977		
Mission: Launch of Voyager 2	August 20	8:29 a.m.
1989		
Mission: Begin Observatory Phase	June 5	—
Mission: Begin Far Encounter Phase	August 6	—
Mission: Begin Near Encounter Phase	August 24	—
Nereid: closest approach	August 24	9:13 p.m.
Ring Plane: crossing inbound	August 24	11:56 p.m.
Rings: Sun occultation	August 25	1:00 a.m.
Rings: Earth occultation	August 25	1:02 a.m.
Neptune: closest approach	August 25	1:02 a.m.
Neptune: Earth occultation	August 25	1:09 a.m.
Neptune: Sun occultation	August 25	1:09 a.m.
Rings: Earth occultation	August 25	1:57 a.m.
Rings: Sun occultation	August 25	1:58 a.m.
Neptune: Earth occultation ends	August 25	1:58 a.m.
Neptune: Sun occultation ends	August 25	1:58 a.m.
Ring Plane: crossing outbound	August 25	2:21 a.m.
Spacecraft Maneuver, Alkaid lock	August 25	2:52 a.m.
Triton: closest approach	August 25	6:16 a.m.
Triton: Earth occultation	August 25	6:45 a.m.
Triton: Sun occultation	August 25	6:46 a.m.
Triton: Earth occultation ends	August 25	6:48 a.m.
Triton: Sun occultation ends	August 25	6:48 a.m.
Mission: Begin Post Encounter Phase	September 11	—
Mission: Encounter period ends	October 2	—
1990		
Mission: Interstellar Phase begins	January 1	—
2000		
Voyager 2: Insufficient power for some instruments	—	—
2018		
Voyager 2: Internal power ends	—	—
2034		
Voyager 2: maneuvering power ends	—	—
2037		
Mission: Telecommunication ends	—	—

Note: Events beyond 2000 are somewhat arbitrary. The final science instrument aboard the spacecraft is expected to be turned off in 2014 because of power limitations. By the year 2025, although maneuvering hydrazine will still be available, there will most likely not be enough power to heat the hydrazine lines and stop them from freezing. The 2034 and 2037 tentative milestones stated in the table might not be reached.

telephoto camera of the imaging system and required two frames to cover the whole disk of the planet and the region of the known rings. The particles and fields instruments started to look for the effects of Neptune's magnetosphere on the solar wind in anticipation of detecting a bow shock produced by the planet's magnetosphere. Additional trajectory correction maneuvers made in this phase "fine-tuned" the path of the spacecraft for an optimum flyby of Neptune and Triton.

The spacecraft traveled closest to Nereid on August 24, at a distance of 2,892,500 miles (4,655,000 km). The near encounter phase began on that same day and continued through August 29. This phase of the encounter produced all the images of greatest value because of their high resolution. One showed the surface of Nereid, a good series of the north polar regions of Neptune, and extreme close-ups of Triton. During this phase, too, the magnetic field of Neptune and the inner magnetosphere were investigated in great detail.

Close to the beginning of the near encounter phase the DSN in Australia transmitted a precise tone at the X-band frequency. The spacecraft received this signal and retransmitted it back to Earth. The returned signal was picked up by the other DSN stations eight hours later. The Doppler shift in the frequency of the returned signal was measured and from it the change in velocity of the spacecraft was calculated. The effects of Neptune's gravity could then be ascertained with improving precision.

All the science instruments began operating in accordance with the preselected schedule. Every six minutes the low-energy charged particle detector collected data about the flow direction of particles in the magnetosphere. The infrared instrument, IRIS, investigated a spot in the southern hemisphere of Neptune where radio signals would later pass through the atmosphere during the occultation experiment. The data gathered would allow the hydrogen/helium ratio to be determined. *Voyager*'s pointing instruments made scans of the sunlit rim of the planet's disk with the cameras of the imaging system and with the photopolarimeter and the infrared instrument. Then the imaging cameras looked at the ring arc region before swinging back to the limb again.

More observations of the rings followed. Starting just under five hours before closest approach the photometer and ultraviolet spectrometer observed the starlight from Sigma Sagittarii as the star appeared to drift behind the rings as seen from the spacecraft. Meanwhile commands were being received from Earth to update the timing of blocks of commands previously recorded at the spacecraft for execution during the encounter. Particularly important was timing for the radio occultation experiment and the image motion compensation at Triton.

Images were taken of still distant Triton, and more images of the rings. The low-energy particle detector switched to sampling higher energy particles as the spacecraft plunged deeper into the magnetosphere. About 55 minutes before closest approach of the spacecraft to Neptune, the various optical devices were pointed away from the planet to avoid their being pitted by particle impacts as the spacecraft crossed the ring plane. The plasma wave experiment was used to count the impacts on the spacecraft of microscopic dust particles in the plane of the rings. Then the spacecraft was rolled so that the instruments could look for charged particles raining into the north magnetic pole of the planet along magnetic field lines. About one hour before closest approach the radio occultation experiment began, and when the signals were received four hours later at Earth they carried information about the rings, the ionosphere, and the planet's atmosphere.

Voyager 2 was directed to hurtle past Neptune at a speed of 60,980 miles per hour (98,350 kph) about 3000 miles (4,850 km) above the planet's cloud tops at a location close to latitude 76 degrees north. This was, in fact, the closest approach to any planet which the *Voyager 2* had flown by since leaving Earth, and it was only 2730 miles (4400 km) above the sensible atmosphere where atmospheric drag would begin to affect the spacecraft. The atmosphere-skimming approach was needed to allow Neptune's gravity to swing the spacecraft close to Triton.

After its closest approach to Neptune the spacecraft passed over the terminator, which divides the sunlit side of the planet from its dark side, and its instruments soon saw the Sun setting into the atmosphere of the planet, an important sequence recorded by the ultraviolet instrument to provide more information about the atmosphere.

With the spacecraft in the shadow of the planet and also hidden from view of Earth. *Voyager*'s radio transmitter's parabolic antenna was programmed to direct its beam automatically so that the radio waves would be made to scan around the limb of the planet from the ingress point in the northern hemisphere to the egress point at 40 degrees latitude in the southern hemisphere. This complicated automated maneuver ensured that despite refraction bending the radio waves as they passed through the atmosphere of Neptune the radio beam would still be directed toward Earth. During the whole of this maneuver other instruments were, of course, still gathering their data about the magnetic fields and particles and obtaining ultraviolet and infrared information about the north polar region of the planet.

Scientists expected that when viewed from the far side of the planet from the Sun, the rings would be enhanced by forward scattered light, possibly to reveal much greater detail than the images obtained by back scattered light during the approach to Neptune. Earthrise and sunrise were viewed from behind the planet, and an edge-on image of the rings was attempted.

After that the spacecraft began to concentrate on Triton. Selecting a suitable path for a Triton encounter called for evaluating several risks; the hazards of zooming in on too close an approach to Neptune, the extent of whose atmosphere was not known with much precision, and the dangers from particles in the ring plane. No one really knew how far the region of ring particles extended from Neptune. Mission planners had to decide how close to the planet they could allow the spacecraft to go skimming over the pole of Neptune while avoiding an extended atmosphere, and what were the chances of a ring plane catastrophe. There was little doubt that Neptune's ring system was different from those of the other planets and that it appeared to be a patchy rather than a continuous system. What was known about ring systems was that they all had thin disks of dusty material inside and most probably outside the visible rings. Such diffuse rings were also probable at Neptune because they had been encountered at the other giant planets. However, *Voyager* had safely passed through the diffuse ring material at Saturn and Uranus.

The probability of *Voyager* passing through an unexpected ring arc was very low, but the problem at Neptune was that the craft had to cross the ring plane before gathering its most important science data. While the spacecraft itself might not suffer damage, its optical instruments could have suffered the effects of impacting dust particles, which would of course have jeopardized that data. This hazard could be avoided by orienting the spacecraft just before it crossed the ring plane and during the crossing so that the optical surfaces were turned away from the line of flight and could not be hit by the particles. Commands to accomplish this orientation had to be stored in the spacecraft in advance of the encounter and automatically executed at precise times.

Another possibility was that perturbations of ring particles by Triton in its highly inclined orbit could give rise to polar rings, as suggested by A. R. Dobrovolskis in 1980 writing in *Icarus.* The *Voyager* spacecraft would travel along a path that would take it close to the plane of any such ring. This was recognized as a potential hazard but one with a very low probability.

Another question was that of the radiation environment, because severe radiation could damage instruments and electronics within the spacecraft. This had, indeed, happened when *Voyager 1* passed close to Jupiter and penetrated the intense radiation environment. That spacecraft had sustained damage to its photopolarimeter, a temporary failure of its ultraviolet spectrometer, and interference with its internal clock, which resulted in timing problems for some of the science experiments. However, the magnetic field of Neptune was believed to be much weaker than that of Jupiter, so that the radiation environment also was expected to be weaker than that of Jupiter. Scientists assumed that Neptune's radiation environment would be similar to that of Uranus, and the maximum exposure to the spacecraft would be almost half that experienced by *Voyager 1* at Jupiter.

An entirely new hazard was posed by the atmosphere of Neptune. Flybys of the other planets had been high above any appreciable atmosphere. Not so with *Voyager* at Neptune. To use the gravity field of Neptune to ensure a close approach to Triton, *Voyager* had to pass inordinately close to Neptune (figure 3.3). In addition, there were other constraints on targeting the flyby. To gather very important data about Neptune and Triton, the path of the spacecraft had to be arranged so that *Voyager* would be occulted by both bodies; there had to be periods of both Sun and Earth occultations at Neptune and Triton.

After several years of soul-searching analysis, mission planners decided that the spacecraft could be targeted at Neptune for a safe passage by the planet and still achieve a flyby of the big satellite at a height of only 23,836 miles (38,360 km) above its expected bizarre surface.

About two hours after closest approach to Neptune the spacecraft was rolled to lock on to a new guide star so that it would be correctly oriented for magnetospheric measurement between the planet and its big satellite while the satellite remained in the sights of the optical instruments.

For eight hours the imaging system, the photopolarimeter, and the infrared spectrometer and radiometer, would be focused on Triton with the expectation of ultimately obtaining images of surface details as small as 0.62 miles (1 km) at the time of closest approach to Triton, just over five hours after *Voyager*'s closest approach to Neptune. Changes in the velocity of the spacecraft because of Triton's gravity would at last allow the mass of the satellite to be ascertained more precisely. A passage of starlight from Beta Canis Majoris through the atmosphere of Triton would be recorded with the photopolarimeter and ultraviolet instruments. The Arabic name of Beta Canis Majoris is Mirzam. It is a bright, magnitude 2, star close to Sirius and slightly below and to the left of the constellation Orion as seen in our Northern Hemisphere winter skies. Next, the spacecraft was oriented for another radio occultation experiment. This time the radio waves would pass through Triton's atmosphere on their way to Earth and provide vital information about that atmosphere. The spacecraft also passed through the magnetotail of Neptune, and the ultraviolet and infrared instruments continued to scan the planet. Auroras and lightning were also searched for on Triton.

Much information was stored in the tape recorders of the spacecraft during the periods of Earth occultation and was sent to Earth after the spacecraft emerged into view from behind Triton. And all the science experiments continued to gather data for several months of the post encounter phase which continued from September 11 until October 2, 1989. During this phase the spacecraft continued to look back on the

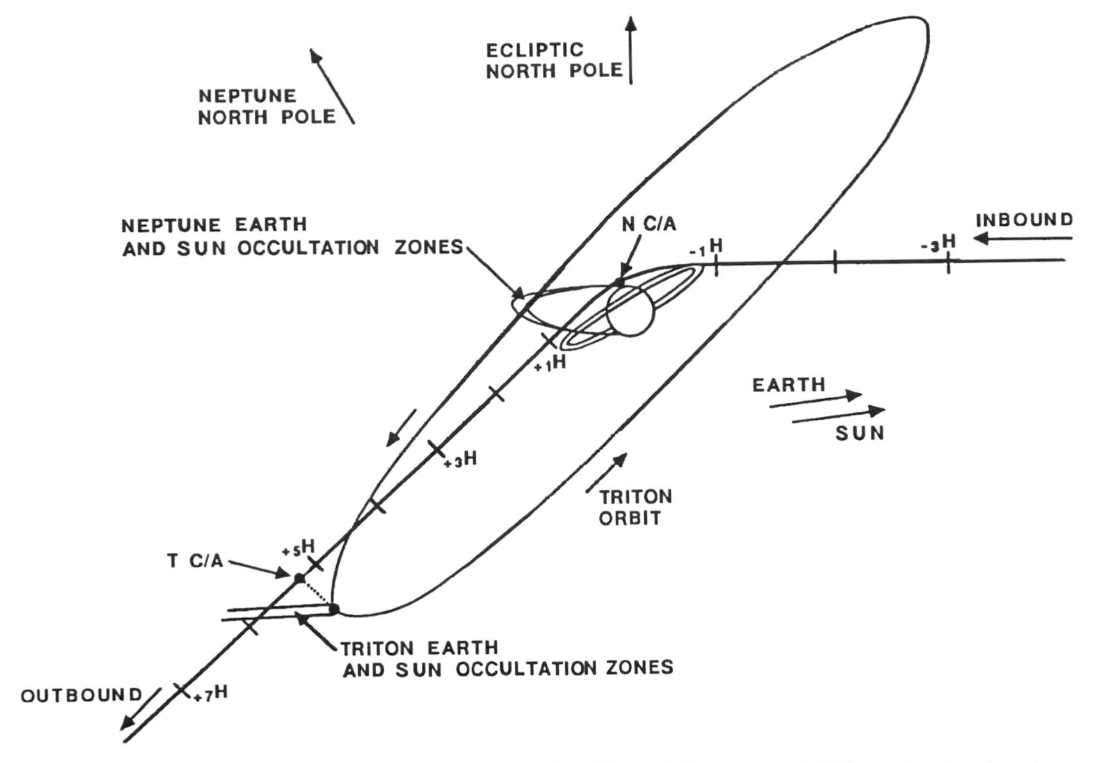

FIGURE 3.3 The proposed path of *Voyager 2* in its flyby of Neptune and Triton showing its close approaches to the rings of Neptune, the planet, and to Triton. Also shown are the occultation zones which the spacecraft should pass through to accomplish its mission. (JPL)

Neptunian system to observe the dark side of the planet and its big satellite and to explore the magnetosphere beyond the planet.

At the encounters with Jupiter and Saturn all the science data, except the imaging data, gathered by *Voyager* were encrypted with an error-correcting code that required as many bits of data for the code as there were for the science information. As the spacecraft moved farther and farther from Earth for its most distant encounters, other ingenious methods had to be applied to return accurately quantities of data over the enormous distances to the outermost planets. A new method of coding was developed that reduced the coding overhead by 15 percent and at the same time reduced the number of errors from 1 in 20,000 to 1 in 1,000,000.

Theoretically, if everything remained constant, the bit rate for return of data should decrease with the square of the distance; this is known as the inverse square law. In other words, if the distance is doubled the data rate is one-quarter, and if the distance is trebled the data rate is one-ninth, and so on. At Jupiter, about a half billion miles from Earth, the highest data rate was 115,200 bits per second, whereas at Neptune, which is 3 billion miles from Earth, the data rate would be expected to be less than 1,500 bits per second. However, the project, through application of high technology of various kinds, managed to ensure a data rate from Neptune of 21,600 bits per second. This was made possible by use of several innovative techniques.

To increase the amount of information that can be reliably returned at the lower data rates, engineers had to devise ways to reduce the number of bits required to transmit science data in general and especially to transmit images which are notorious for demanding numerous bits. These ways included changing the way the spacecraft encodes the data bits before sending them, and editing and compressing data by various techniques. For the Uranus encounter, all science data, including imaging data, was coded in a way to reduce the overhead to about 15 percent. This same coding was used at Neptune, which, with upgrading of the DSN, allowed data to be returned at the 21,600 bits per second rate mentioned above; slightly less than the bit rate from Uranus.

First used very effectively during the encounter of *Voyager* with Uranus, a method of compressing imaging data was again used at Neptune. This method reduces the number of bits for each image. An imaging frame consists of 800 lines each containing 800 pixels, or picture elements; that is, a total of 640,000 pixels per image. Eight bits of data are needed to express 256 levels of grey (from white to black) with the consequence that a single image requires 5.12 million bits of data. Transmitting such an image at 21,600 bits per second would take almost 4 minutes. So that image data could be compressed each line of pixels was divided into blocks of five pixels. The absolute brightness of the first pixel in each line was sent, and then the brightness of each of the following pixels was expressed as its difference, if any, from the brightness of the preceding pixel. Because much of a typical image consists of areas of uniform grey scale, for example, areas of darkness or of uniform brightness, this method effectively discards unnecessary picture elements and reduces by as much as 60 or 70 percent the number of bits needed to be transmitted for each image.

A typical compressed image of 2 million bits takes about 1½ minutes to transmit. The radio waves carry the image through space at 186,000 miles per second. During the times it takes to transmit the whole of the image, the radio waves carrying the beginning of the sequence of signals travels a distance of 17 million miles toward Earth before the last bit of the image leaves the spacecraft's transmitter. Each image thus speeds through space like an arrow of information whose shaft is 17 million miles in length. If images were sent immediately one after the other, more than 170 images could be strung out in space over the almost 3 billion miles from Neptune to Earth. Actually, because other data had also to be returned to Earth in real time, the telecommunications facilities allowed the receipt of only about 12 to 17 images per hour and an average of about 200 images per day from *Voyager* during the encounter with Neptune.

In 1981 the scan platform of *Voyager 2* jammed in one axis following closest approach to the planet Saturn. This problem limited the operation of the scan platform for the remainder of the encounter sequence at that planet. Two days later the platform could again be moved. A long period of ground testing and analysis followed through which engineers discovered that the problem had arisen from a loss of lubricant caused by the scan platform's having been operated at high speeds for too long a time. The high-speed gear train of the platform became inoperative because of a shortage of lubricant. However, after the platform was rested for a time, the lubricant seeped back into the gear train and the platform could again be operated. But its operation had to be restricted to low speed scans. This applied throughout the encounter with Uranus and during the cruise to Neptune. Controllers found that if the platform continued to be operated at low speeds only, it could be expected to remain serviceable during the Neptune encounter also.

In 1978, during April and before *Voyager 2* reached Jupiter, the spacecraft had automatically switched to its backup receiver; the primary receiver had failed. This backup receiver unfortunately had a problem in its tracking loop capacitor which prevented the receiver from locking onto a changing uplink frequency. The change in frequency is normal. It results from the effects of Doppler shift arising from changes in the velocity between the spacecraft and the Earth. A switch back to the primary receiver did not work and the mission had to rely upon the ailing backup receiver. During the long cruise periods between planetary encounters engineers determined how the tuning of the spacecraft's receiver depends on temperature and how this, in turn, is affected by the operation of the various subsystems within the spacecraft, all of which develop waste heat. To ensure that the receiver would have time to stabilize its temperature after operation of the spacecraft's subsystems, commands were not sent to the spacecraft until enough time elapsed for the receiver's temperature to stabilize. Another backup capability was to send commands to the spacecraft at slightly different frequencies to ensure that one set of commands would be picked up by the receiver. As mentioned earlier, as a contingency against the receiver failing completely, a backup computer on board the spacecraft was loaded with simplified encounter routines which would be automatically implemented and the results sent back to Earth even if the spacecraft could no longer be contacted from Earth.

Several of the contingency techniques were first employed at Jupiter encounter and refined for subsequent encounters with the other outer planets.

At Uranus, *Voyager* encountered the problem of low light levels when imaging the Uranian system. The intensity of sunlight follows the inverse square law, which results in the sunlight at Neptune being about 900 times less intense than sunlight at Earth. This meant, of course, that longer camera exposures had to be given at Neptune than at Jupiter, Saturn, or Uranus. The problem was that the spacecraft, Neptune, and its satellites were all in relative motion. Most photographers know the problem of trying to get clear pictures of moving objects when low light conditions make long exposures essential.

To be sure of obtaining good pictures of Triton's surface at high resolution and of the elusive rings of Neptune despite the high speed of the spacecraft and the extremely feeble levels of lighting in the Neptunian system, innovative techniques were required to control the orientation of the spacecraft and to operate its camera systems. To avoid smeared images, the imaging team used a technique of image compensation that had first been applied at Saturn during the *Voyager* mission. This technique relies upon moving the camera during the exposure to compensate for the movement of the object being imaged. There are several ways to do this.

Classical image motion compensation was used at Uranus. It involved using not the motion of the ailing scan platform but *Voyager*'s thrusters to rotate the whole spacecraft. That way, its instruments could track the target during exposure of each image. The problem in using the classical method is that the movement of the spacecraft shifts its antennas away from being pointed to Earth, thereby breaking communications. The images have to be stored in the spacecraft's tape recorders until after the imaging sequence when the spacecraft's antenna can be turned back toward Earth again.

Another method of image-motion compensation that was developed for the encounter with Neptune is to rotate the spacecraft only to a point close to where further movement of the antenna would cause communications to be interrupted. Then the spacecraft rotates back for the next image. Images acquired with this nodding image-motion compensation can be transmitted as they are taken, because the spacecraft never loses communication with the terrestrial ground stations.

A maneuverless image-motion compensation technique uses only the movable scan platform on which the cameras are mounted. The attitude of the spacecraft remains unchanged while the platform is turned in elevation only and at a low scan rate to track the target. Communications to Earth are maintained with this method also. This method, too, was used at times when appropriate because high slew rates were not needed.

To make the spacecraft a more stable platform for the long exposure imaging, personnel of the Voyager Flight Teams at the Jet Propulsion Laboratory compiled new computer software for the *Voyager.* This ensured that whenever the spacecraft's tape recorder started or stopped the onboard computer would command the spacecraft to fire its thrusters to counter changes in attitude caused by reaction to the operation of the tape recorders. In addition, the duration of firing of the attitude control thrusters was reduced from 10 milliseconds to 4 milliseconds to ensure a steadier spacecraft by reducing the changes to attitude of the spacecraft by each firing of the thrusters.

At the final press conference following the encounter of *Voyager* with Neptune, the last of the 54 press conferences held during *Voyager*'s 12-year epic journey through the outer Solar System, Norman Haynes, the current project manager, emphasized how everything had worked perfectly and how the flight teams had fully utilized 99 percent of the capabilities of the spacecraft. The ailing receiver gave no trouble; all commands to the spacecraft were received successfully. The "rheumatic" scan platform performed admirably without any hitches. The spacecraft came within 20 miles (30 km) of the aim point at Neptune. Nearly 80 spacecraft maneuvers were performed automatically during the flight through the Neptunian system.

On Earth a prodigious technical feat was performed in arraying the 30 antennas of the VLA and Goldstone facilities to receive signals from the spacecraft simultaneously and to bring them all into synchronism so that they could be added together.

The Air Force's Positioning Satellite System was also used to coordinate time keeping of the antenna systems around the world. In California, the Southern California Edison utility company developed a *Voyager* emergency backup plan with personnel on 24-hour duty to ensure that there could be no loss of power at the DSN station for tracking and commanding the *Voyager* and receipt of data from the spacecraft.

At the same press conference Bradford Smith, Imaging Team leader, stated that the epic mission of *Voyager* was "a very special decade in human history" that will be remembered as the age of the exploration of the outer Solar System in the same way that Magellan is remembered as the first explorer to circumnavigate the Earth.

Voyager's encounter with Neptune ended officially on October 2, 1989. It resolved a number of important questions about the Neptunian system, said Edward Stone, Project Scientist. The first key question was whether Neptune possessed a magnetic field. *Voyager* supplied an affirmative answer, but surprised everyone by showing that the field was tilted and offset similar to that of Uranus despite the different spin axes of the two planets.

The second question was about the speed of rotation of Neptune. When *Voyager* discovered radio emissions with a periodicity of 16 hours 3 minutes the question was answered. This was the Neptunian day for the bulk of the planet. Neptune thus rotates slightly faster than Uranus but much slower than Jupiter and Saturn.

The third key question was about weather systems on Neptune. The results from *Voyager* provided another set of surprises. Despite the remoteness of Neptune from the Sun the planet's atmosphere is surprisingly active. It has a Great Dark Spot somewhere analogous to the Great Red Spot of Jupiter, high cirrus type clouds, retrograde winds peaking at 700 miles per hour (1100 kph), and a latitude temperature profile similar to that of Uranus.

Did Neptune have small satellites other than Nereid? Again *Voyager* provided an answer. Six additional satellites were discovered, all with dark surfaces, all irregularly shaped, and none showing geological modification that might be comparable with that on the small Uranian satellite, Miranda.

What about the ring arcs? Did they exist? *Voyager* showed that there are, indeed, ring arcs at Neptune but that they are part of a complete ring. The spacecraft also discovered the planet had other rings.

Another question concerned the size of Triton, the nature of its atmosphere, and what type of processes might be occurring on its surface. The radius of the big satellite is 840 miles (1350 km). Its density is just over twice that of water, which suggests it contains more rocky material than the icy satellites of some of the other outer planets. Triton seems more akin to Pluto. The atmosphere is predominantly nitrogen and methane but at a very low pressure. The satellite is colder than was expected and the *Voyager* images show that it possesses a wide diversity of geologic features.

The results from *Voyager* provide an encyclopedia of the outer Solar System for decades to come, asserted Stone. And its journey will continue and will hopefully be monitored for another 25 years searching for the edge of the Sun's influence, he concluded.

Subsequently the spacecraft began to travel out from the Solar System joining *Voyager 1* and *Pioneers 10* and *11* in searching for the heliopause. The spacecraft continued examining radiation from ultraviolet stars and detecting fields, particles and waves in the space of the outer Solar System beyond all the known planets. Meanwhile, *Voyager 1*'s cameras were reactivated to make a unique group picture of the planets of the Solar System (figure 3.4) early in 1990 from its viewpoint far beyond Neptune. In this imaging sequence all of our planets, including the Earth and its billions of people, were rapidly becoming lost in the glare of a mediocre yellowish star. On both *Voyager 1* and *Voyager 2* some instruments continued to observe the distant regions of space searching for the boundary of the Solar System.

The heliopause is the boundary between the "bubble" in space, which is dominated by the solar wind, and the plasma flows of interstellar space. Scientists expect to discover a bow shock where the supersonic solar wind is slowed and bent around to form a tail away from the Solar System, stretching like a gigantic windsock down the interstellar wind. Although the *Pioneers* first traveled into the outer Solar System, the *Voyagers* may be the first to reach the heliopause because they are traveling faster than the *Pioneers*. It may, however, be some 10 to 30 years before either spacecraft crosses the heliopause and emerges into interstellar space. It is estimated that the shock will not be encountered until the spacecraft are at least 61 astronomical units from the Sun. Researchers at the Jet Propulsion Laboratory expect that communications with the

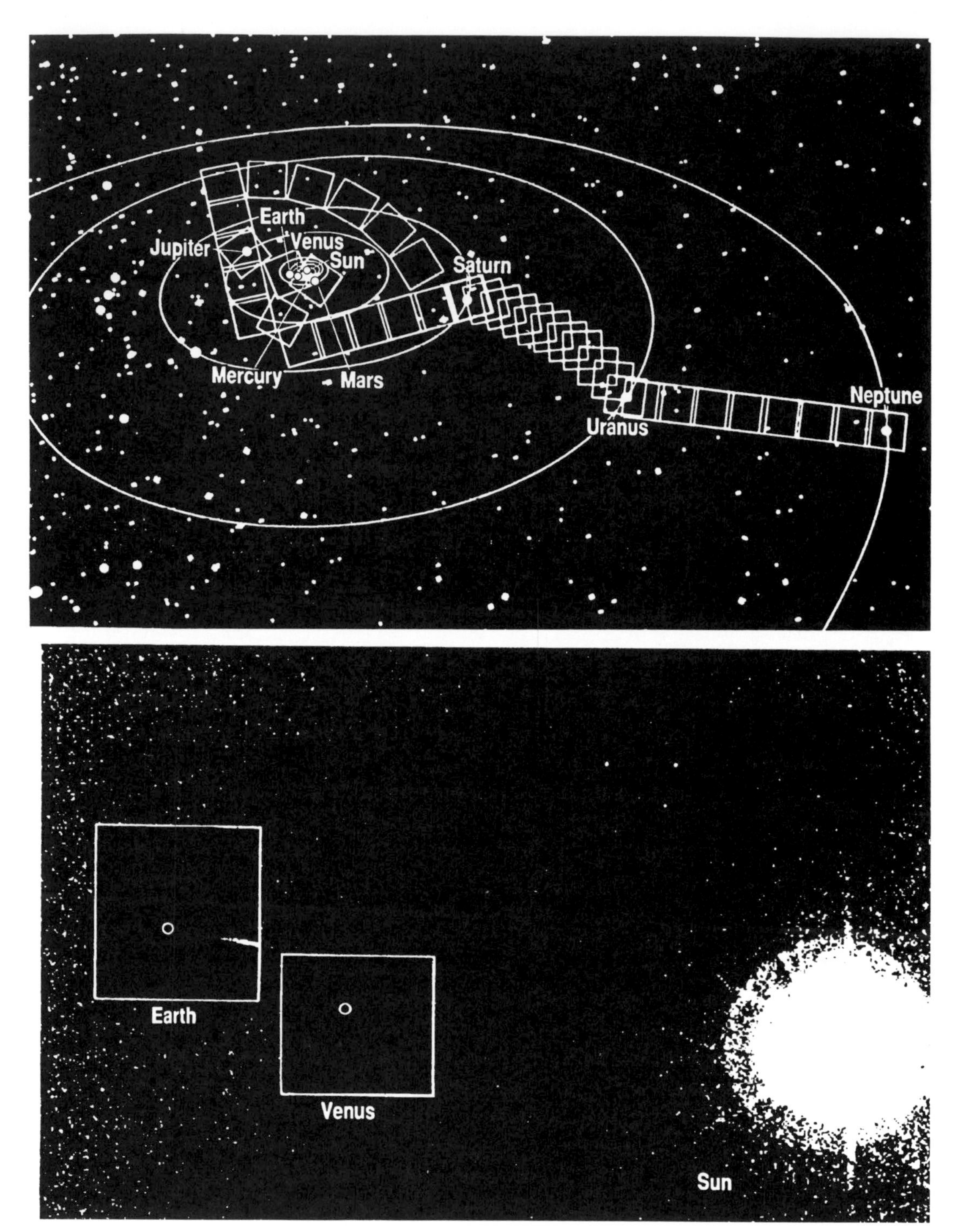

FIGURE 3.4 *Voyager 1* took this series of parting shots at the Solar System and its planets as the spacecraft traveled out toward the distant stars. A series of images were taken across the Solar System to include pictures of the Earth, Venus, Jupiter, Saturn, Uranus, and Neptune as identified by the white rectangles on (A). The lower illustration (B) shows the Sun and the tiny crescents of Earth and Venus (within the white circles inside the white rectangles) almost hidden in the glare of the Sun which itself appears as a bright star just east of Orion's belt. We see our Earth in a realistic perspective, a mere dot of light lost amidst other brighter dots of light. How futile seem our earthly struggles for fame or for wealth, for territories and for ideological political systems on this insignificant spot of light viewed against such a small part of the cosmic backdrop. And yet how daring and inspiring are our attempts to reach and try to understand the frontiers of our Solar System. (NASA/JPL)

Voyagers will continue until about 2020, until a time when the electrical power from the RTGs will become too low to keep the spacecraft operating.

Even when the spacecraft has lost all communication with Earth and it continues toward the stars as a lifeless machine it will have an important purpose: bearing a message from Earth to any extraterrestrial intelligence that may be lucky enough or smart enough to intercept its inert hulk. Together with the *Pioneer* spacecraft and *Voyager 1*, *Voyager 2* carries a message to extraterrestrials.

When *Pioneer 10* was being tested in a space simulator at TRW systems, Redondo Beach, California, the author, who was then a natural science correspondent of the *Christian Science Monitor,* realized that when the spacecraft swung past Jupiter it would have sufficient momentum to escape from the Solar System. A few evenings previously he had visited with Merton Davis of the Rand Corporation who was entertaining a mutual friend, the late Arthur C. Cross who with the author had earlier been very active in the British interplanetary movement. Also at the gathering at Davis's home in Santa Monica was Carl Sagan. He told of how he and Frank Drake had recently attended a science conference in the Crimea discussing how mankind might communicate with extraterrestrials and send a message that might be intercepted.

The author realized that *Pioneer* presented such an opportunity. He mentioned the idea to Don Bane, then with the *Los Angeles Herald/Examiner,* and to Richard Hoagland a freelance writer. They were enthusiastic and agreed with the author that any plaque on the *Pioneer* should not be a message from congress or the president as had been done with the plaques on the Moon. Instead it should be a message from humankind. The author talked with Sagan that same day and he agreed to approach NASA to place a message on the *Pioneers.* Later the author was on a panel in Boston with Robert Cowen, science editor of the *Christian Science Monitor,* and Homer Newell, then an assistant administrator at NASA. Newell mentioned to the author over lunch that his message would be on the spacecraft. It had been formulated by Sagan and Drake with artwork designed by Linda Salzman Sagan who was at that time Sagan's wife. Newell explained to the author that nothing should be released in the press until the launching. So secrecy prevailed. Indeed, this proved very wise because when the plaque and its design were released a great outcry arose that it was racial, pornographic, or should not have been sent for various reasons.

However, the author stated in a NASA publication (SP-446) that the plaque "signifies an attribute of mankind that, in an era when troubles of war, pollution, clashing ideologies, and serious social and supply problems plague them, mankind can still think beyond themselves and have the vision to send a message through space and time to contact intelligence in a star system that perhaps has not yet condensed from a galactic nebula."

"The plaque," I wrote, "represents at least one intellectual cave painting, a mark of Man, that might survive not only all the caves of Earth, but also the Solar System itself. It is an interstellar stela that shows mankind possesses a spiritual insight that transcends the material problems of the age of human emergence into space."

The idea continued into *Voyager,* for which Sagan and some associates designed a message of much greater complexity than the initial message on *Pioneer.* Each *Voyager* carries a gold-plated record containing 116 pictures of terrestrial objects and people and 90 minutes of sound recordings of music, human voices, and sounds from Earth. Whether or not such a complex message will be intelligible to other intelligent but alien creatures, should any exist, is a moot point. Even on the original simple *Pioneer* plaque there were doubts about illustrative material such as whether an arrowhead indicating motion in a certain direction would be intelligible to a being that had not gone through a phase of using bows and arrows. Nevertheless the idea of any message to beings elsewhere is a powerful expression of humanity's desire to be known and remembered. Like messages in bottles cast on the oceans of Earth, the four messages may never be intercepted among the stars, just as many prehistoric cave paintings have probably never been seen by mortal eyes other than their originators. But a billion years from now, without any catastrophic encounter in space, the derelicts from Earth may be intercepted and interpreted. That, indeed, may give comforting and horizon-expanding thoughts to some creature incredibly distant in space and time who, looking out with wonder into the cosmos, will realize with a burgeoning curiosity that it is not the only form of intelligent life to have been created.

4

THE PLANET NEPTUNE

Some of the images sent to Earth from *Voyager* as it approached Neptune in May 1988 began to show the planet with a resolution greater than that obtainable with Earth-based telescopes. An image captured on May 9 from a distance of 426 million miles (685 million km) showed both Neptune and Triton, but then Triton was still only a point of light occupying one picture element (pixel) while Neptune had a diameter of eight pixels. Meanwhile, on July 14, 1988, a ground-based series of images taken by Heidi Hammel of Jet Propulsion Laboratory over a period of five-and-a-half hours at Mauna Kea Observatory, Hawaii, clearly showed the movement of a bright feature, probably consisting of clouds, across the face of the planet. This confirmed observations made in earlier years from Earth. Neptune, in contrast to Uranus, shows distinct atmospheric features. The more significant of these earlier observations were some by Bradford Smith and Richard Terrile in 1983 at Las Companas Observatory, Chile, using a charged-coupled amplifier with a methane filter. This type of amplifier is an advanced electronic detector capable of producing images of very faint objects.

As *Voyager* approached closer to Neptune, details became clearer on images the spacecraft sent to Earth. A bright cloud feature, similar to but not necessarily the same as that recorded by Heidi Hammel, was conspicuous on images returned on January 23, 1989. It was at 30 degrees south latitude. The image was obtained at a distance of 185 million miles (309 million km) from Neptune. The imaging team combined images obtained through different color filters to show more subtle details such as a dark band encircling the south pole. The planet had a circumpolar banded appearance similar to that seen on each of the other giant planets. That distinct features could be seen so far from the planet augured well for the possibilities of seeing at Neptune much more intricate cloud features than those present on bland Uranus. However, abundant polar hazes were absent in the polarimetry data, as they were at Uranus. This was in stark contrast to conditions observed at Jupiter and Saturn. With the polarimeter Neptune appears quite bland, like Uranus, when observed in ultraviolet radiation at 2700 angstroms, but in contrast to Uranus shows distinct albedo differences at 7400 angstroms in the near infrared region of the spectrum, notably the bright feature which has been observed from Earth.

A major discovery was made in the Spring of 1989 when images taken on April 3 at a distance of 129 million miles (208 million km) were returned to Earth. Neptune had a large, dark, oval spot, located about 25 degrees south latitude, intriguingly similar to the conspicuous and long-lasting Great Red Spot of Jupiter, which is at about the same latitude on that planet. Relative to the diameter of Neptune, the spot was as large as that on Jupiter compared with the size of Jupiter. The Great Dark Spot on Neptune extended from 20 to 30 degrees south latitude and over 35 degrees in longitude, which represents a length of about 9300 miles (15,000 km). By May, 1989, the images showed a very dynamic atmosphere with atmospheric bands and bright and dark spots. The dark oval was embedded in a dusky belt at its latitude, and the southern edge of the oval had an arc of brightness on its border. Subsequently, features at different latitudes were seen to be moving around the planet at different rates.

As the spacecraft continued toward Neptune, high resolution images showed increasingly complex cloud patterns and features as small as a few hundred miles in extent. A complex atmospheric structure was revealed by these images (figure 4.1) as the spacecraft hurtled toward its closest approach. Such an active atmosphere had not been expected at Neptune because of the meager amount of solar energy received by the planet: one thousandth of that received by Earth. The dusky band surrounding the south pole consisted of a double ring. Details of the Great Dark Spot became clearer and a smaller dark spot closer to the pole was also found. Bright wispy clouds were seen over the Great Dark Spot at its southern edge (figure 4.2). These cirrus-like clouds formed the bright arc seen on earlier images of less resolution. They produce a bright feature on the images which was referred to as the bright companion to the Great Dark Spot. Later, ground-based images taken by H. Hammel at the Mauna Kea Observatory, Hawaii, at the same time as spacecraft images, were compared with the images from the spacecraft. They showed conclusively that the bright area seen on ground-based images from which a period of rotation had been calculated for Neptune was a fuzzy representation of the bright companion of the Great Dark Spot revealed in detail on the *Voyager* images.

Other cirrus clouds formed streaks across the northwest boundary of the Great Dark Spot. These bright clouds were clearly higher in the atmosphere than the dark spot. An elongated dark feature developed over the relatively short period of three planetary revolutions, stretching toward the equator from the west edge of the spot in a northwesterly direction for nearly 10,000 miles (16,000 km). While the feature was darker in shade than surrounding areas, researchers were not clear whether its material flowed from the Great Dark Spot or its lower albedo resulted from an atmospheric disturbance.

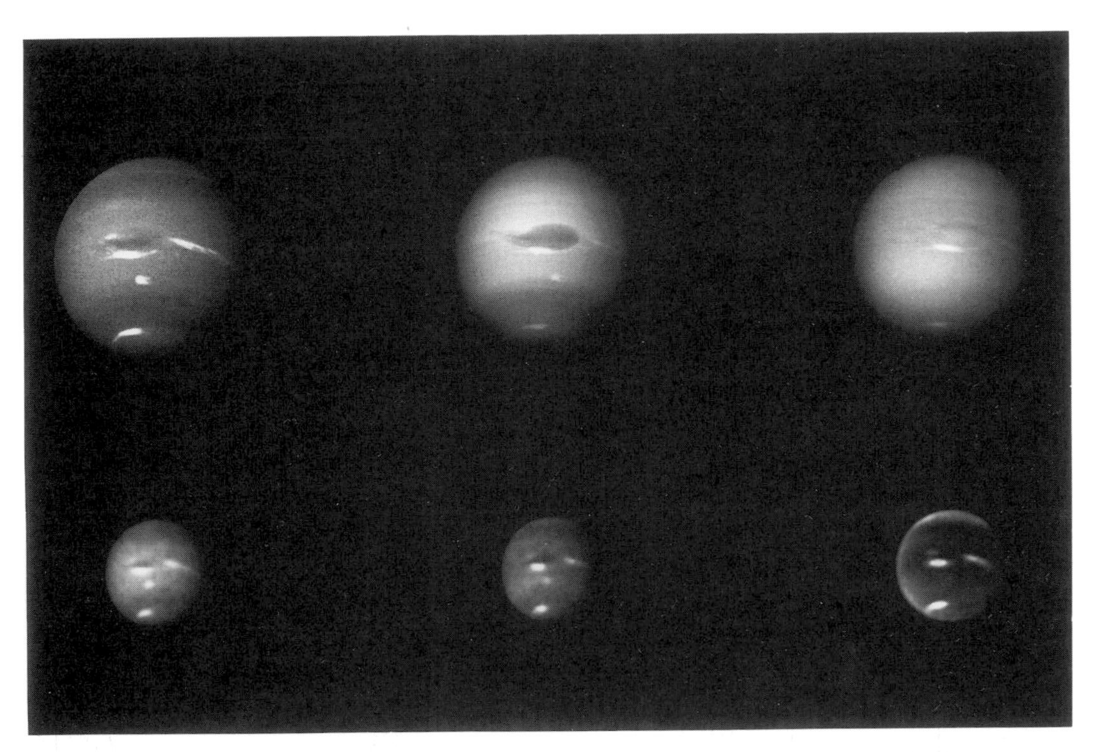

FIGURE 4.1 This series of six images taken through different filters reveals altitude differences in Neptune's clouds. The top three images were obtained through orange, violet, and ultraviolet filters. They show several bright features. Note that the "Scooter" just below the Great Dark Spot is quite clear in the orange image but is not visible in the ultraviolet image. This implies that the "Scooter" is lower than other bright clouds and has more ultraviolet scattering atmosphere above it. Note, too, that the contrast of the Great Dark Spot is reduced in the ultraviolet image. It also is a low feature. The bottom three images were obtained with filters that isolate the effects of methane in the atmosphere, with most absorption by methane in the righthand image. Again the loss of the "Scooter" in the righthand image indicates that there is more methane above this feature, implying that it is deeper in the atmosphere. The images were obtained when *Voyager* was 9.9 million miles (14 million km) from Neptune. (NASA/JPL)

Later this feature was observed to break into a string of small dark spots during five Neptune days (figure 4.3, p. 63). False color images showed methane hazes along the limb of the planet and some haze over the south polar region.

Soon it became apparent that in some regions of Neptune the atmosphere is as dynamic and variable as the terrestrial atmosphere, but on an enormously larger scale. The Earth, for example, would just about fit nicely into the Great Dark Spot. Cloud features moving at the periphery of the Great Dark Spot indicate that the material of the spot rotates counterclockwise similar to the flow around Jupiter's Great Red Spot. Both the big spots are anticyclonic. Neptune's spot appeared during the *Voyager* encounter to be more variable in size and shape than Jupiter's large spot. The Great Dark Spot of Neptune also exhibits spiral arms, and sections of its borders are marked by bright cirrus clouds (figure 4.4, p. 64). The photometric experiments at 7500 angstroms revealed that the bright companion to the Great Dark Spot is at a higher altitude in the atmosphere than the dark spot.

Whereas cirrus clouds in Earth's atmosphere consist of water ice crystals, those in Neptune's atmosphere are of frozen methane. These bright cirrus clouds changed

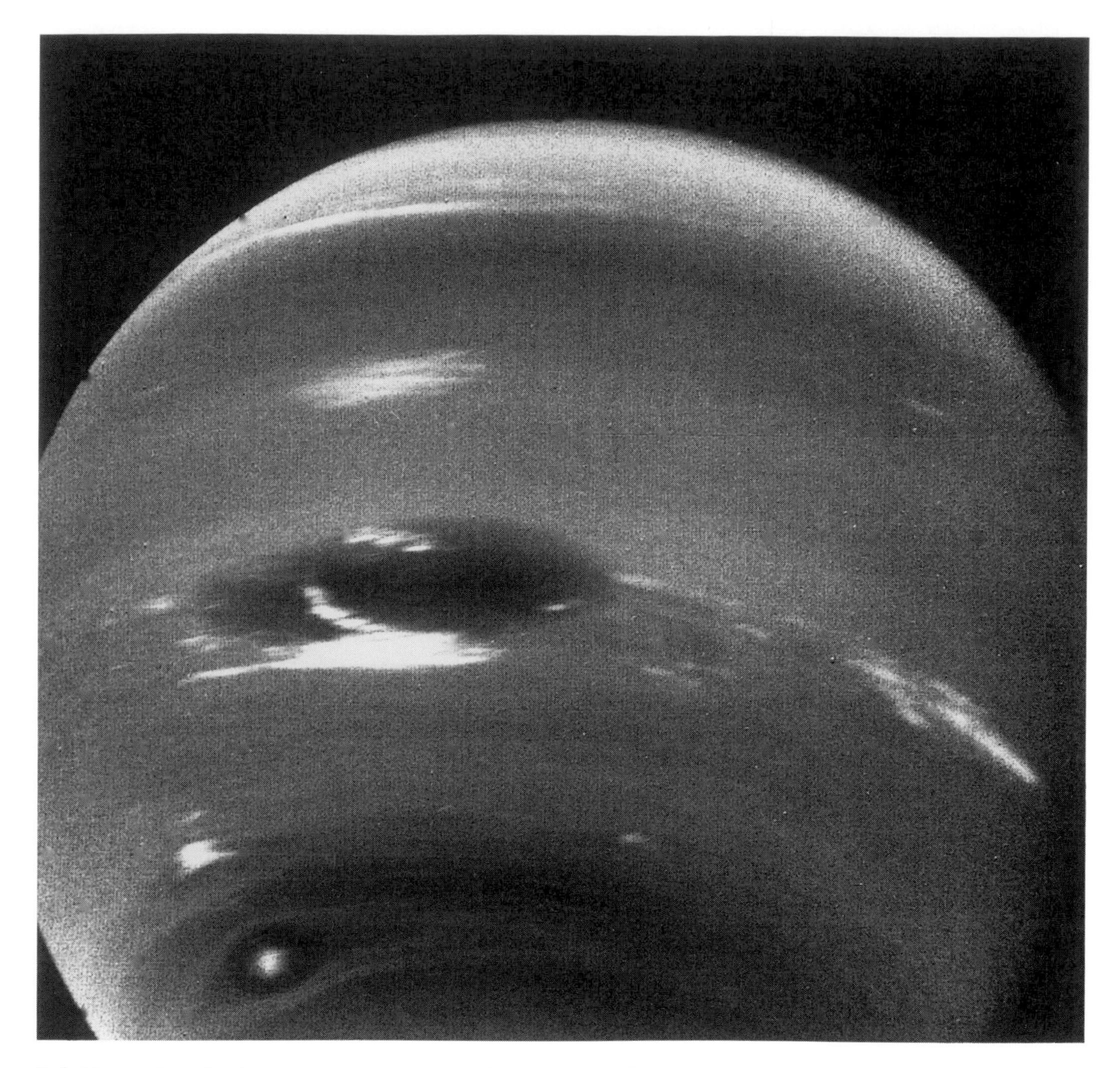

FIGURE 4.2 The Great Dark Spot with its bright companion is slightly to the left of center of this image taken at 3.8 million miles (6.1 million km) from Neptune. The small bright "Scooter" is below the dark spot, and a second dark spot with a bright central area is below the "Scooter." Strong winds at up to 400 miles per hour (640 kph) cause the second dark spot to overtake and pass the larger dark spot every five days. (NASA/JPL)

rapidly during the encounter, especially around the edges of the Great Dark Spot. But it also became apparent that the clouds were forming and dissipating as atmospheric mass rose and fell at the boundaries of the spot. This is somewhat analogous to the formation of lenticular wave clouds located over terrestrial mountains as warm moist air is forced upwards to regions with a low enough temperature for the volatile (water) to condense and become visible as clouds. While the clouds remain almost stationary (figure 4.5, p. 65), the winds blowing through them can be moving at very high speeds.

Voyager also discovered clouds around north latitude 27 degrees that were casting shadows on lower cloud decks estimated as 30 to 60 miles (50 to 100 km) below the cloud streaks (figure 4.6, p. 66). The clouds ranged in width from 30 to 125 miles (50 to 200 km), but the widths of the shadows were much less. In the south polar regions

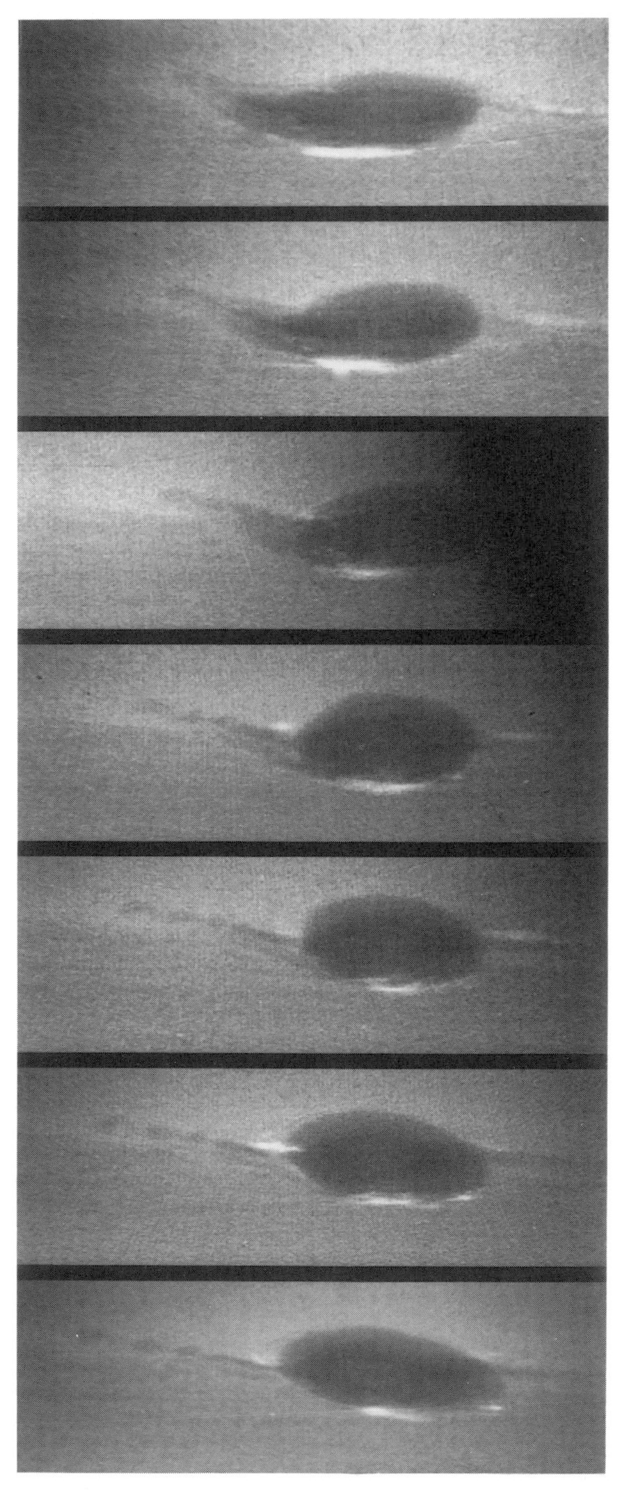

FIGURE 4.3 These images show changes in the clouds around Neptune's Great Dark Spot over a four-and-a-half day period. From top to bottom the images show successive rotations of the planet; an interval of about 18 hours at the 22 degrees latitude of the spot. The images were taken through a violet filter by the narrow angle (telephoto) camera. The sequence shows a large change in the western end (left side) of the spot. A dark extension apparent in the first images converges into a string of small dark spots. The apparent motion of smaller clouds at the periphery of the spot suggests a counterclockwise rotation of the spot, similar to the rotation of Jupiter's Great Red Spot. That such a large rotating spot should be present on Neptune similar to the big spot on Jupiter is surprising in view of the fact that the total energy flux from the interior of Neptune and from the Sun is only five percent of the total energy flux at Jupiter. (NASA/JPL)

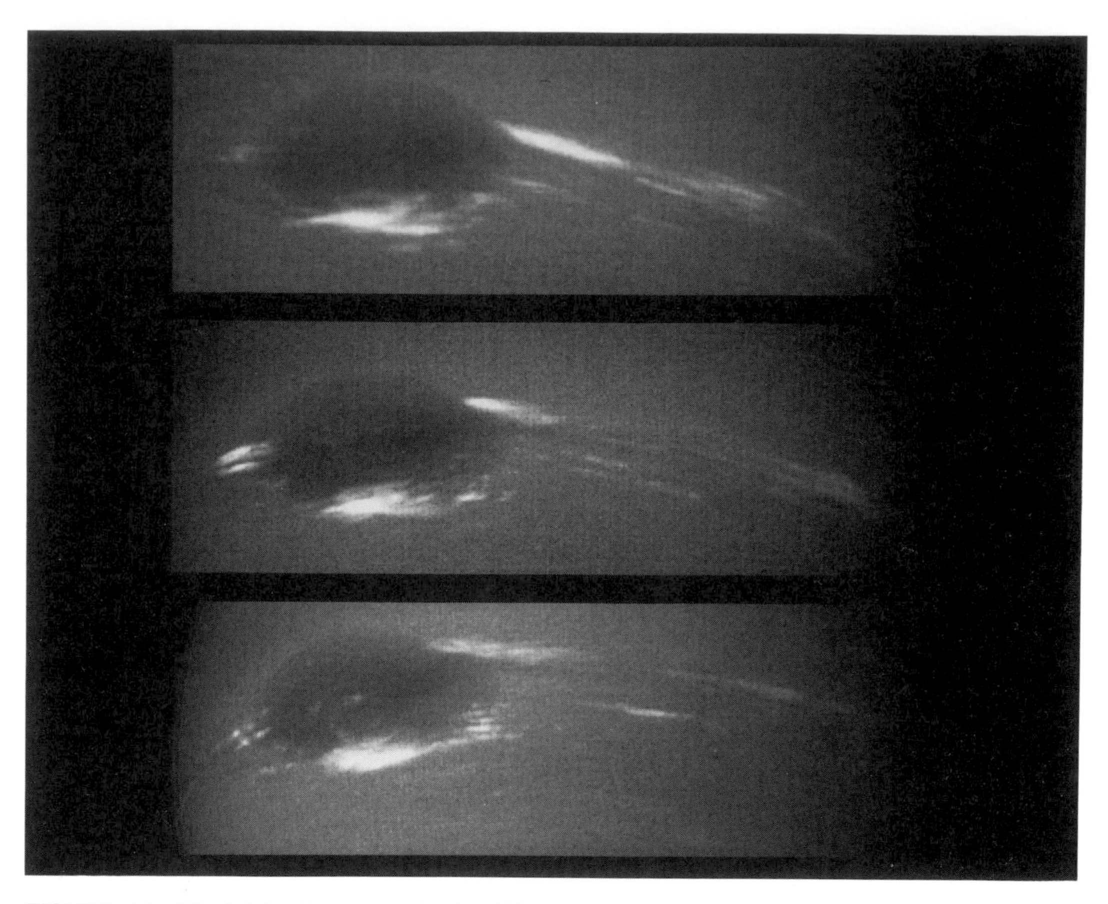

FIGURE 4.4 The bright cirrus type clouds of Neptune change rapidly, often forming or dissipating in a few hours. This sequence spanning two rotations of the planet shows rapid changes of a dynamic atmosphere on an immense scale compared with the dynamics of the terrestrial atmosphere. The Earth is about the size of the Great Dark Spot. The cirrus clouds on Neptune are composed of frozen methane in contrast to the water ice crystals of terrestrial cirrus clouds. (NASA/JPL)

close to the terminator similar clouds cast distinct shadows on lower cloud strata (figure 4.7, p. 67). From the angle of sunlight on the clouds, scientists calculated that these floated 30 miles (50 km) above the lower cloud deck (figure 4.8, p. 68). These high clouds also are believed to be formed of methane crystals. Clouds of this type had not been observed on any other of the giant planets. Haze layers, thought to consist of particles produced by solar radiation acting photochemically on methane, were observed above the cirrus-type clouds.

A small bright spot closer to the pole than the Great Dark Spot moved around Neptune at a faster rate and was quickly nicknamed "Scooter." Time-lapse type motion pictures produced from a series of images showed that the Great Dark Spot appeared to oscillate and the shape of the Scooter changed during the encounter. This bright spot (figure 4.9, p. 69), at about 42 degrees, moved around the planet in 16.1 hours, the same as the rotation period of the planet as derived from the rotation of the magnetic field, and two hours less than the Great Dark Spot. The structure of the spot changed from rotation to rotation and appeared to be breaking down into a series of gravity waves somehow focussed on the spot. The spot is a deeper blue than the rest of the planet, but how far the structure of the spot extends below the visible clouds is uncertain. Some theorists are of the opinion that these big spots extend deep into the atmosphere of the big planets; others that they are phenomena confined to the cloud layers.

At increased resolution intricate structure could be seen within the "Scooter," making it appear somewhat analogous to a terrestrial hurricane seen from orbit. Farther south, at about 55 degrees latitude, another but smaller spot, dark but with a bright central region, appeared on the *Voyager* images (figure 4.10, p. 70). Increases in resolution of the images as the spacecraft bore down on the planet revealed that the central region of this spot consisted of swirling clouds indicative of upwelling and rotation about the center. Bands around the spot suggested the presence of strong winds. It appeared to be a massive storm system.

Overall, Neptune emits 2.8 times as much heat as it receives from the Sun. *Voyager's* instruments measured the amount of heat radiated by the planet's atmosphere from which, together with information from other experiments, the temperature could be derived. Stratospheric temperatures were 750 K (900° F) whereas lower in the atmosphere at the 100 millibars level temperatures were much less, around 55 K (−360 deg. F).

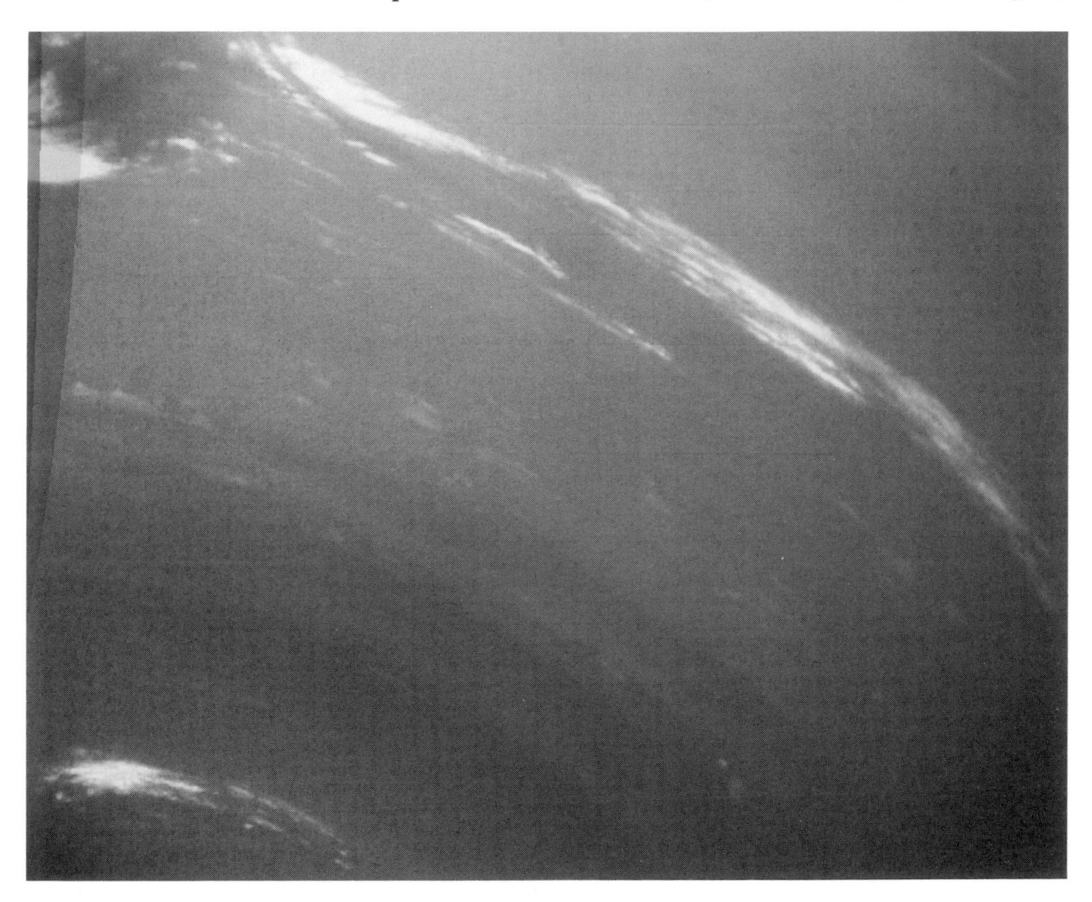

FIGURE 4.5 This image was taken when *Voyager* was 370,000 miles (590,000 km) from Neptune. Structure of clouds in the dark regions near the pole as well as the bright clouds east of the Great Dark Spot is revealed in this specially processed image. Small trails of similar clouds trending east to west and large-scale structure east of the Great Dark Spot all suggest that waves are present in the atmosphere and play a major role in the type of clouds that are visible. (NASA/JPL)

FIGURE 4.6 Taken two hours before the spacecraft's closest approach to Neptune, this image shows vertical relief in the cloud streaks. The linear cloud forms are stretched approximately along lines of constant latitude, and the Sun is toward the lower left. The bright sides of the clouds which face the Sun are brighter than the surrounding cloud deck because they are more directly exposed to the Sun. Shadows can be seen on the side opposite the Sun. These shadows are less distinct at short ultraviolet wavelengths because they are cast on a lower cloud deck and scattering of light by the atmosphere above them diffuses light into the shadow. The shadows are darkest when observed in red light because molecules scatter the longer waves of red light less than blue or ultraviolet light. The widths of the cloud streaks range from 30 to 125 miles (50 to 200 km) and their heights appear to be about 30 miles (50 km) above the main cloud deck. (NASA/JPL)

The temperature curves derived from the infrared instrument revealed much thermal structure and gradients at different levels of the atmosphere with isolated hot and cold locations. Temperatures as a function of height, actually different pressure levels, in Neptune's atmosphere are compared with those of Uranus in figure 4.11, p. 71.

Scientists discovered that the atmosphere above the clouds is hotter at the equator and the poles than at mid-latitudes (figure 4.12, p. 72). Somewhat similar to those on Uranus, temperatures on Neptune are almost the same at the poles and equator while several degrees lower at mid-latitudes. This occurs even though Uranus presents its poles toward the Sun whereas Neptune presents more nearly equatorial regions. In addition, as mentioned earlier, Neptune emits 2.8 times as much energy as it receives from the Sun compared with Uranus, which emits a much smaller amount (only 12 to 14 percent) of the incident solar radiation. Circulation may be controlled by heat dis-

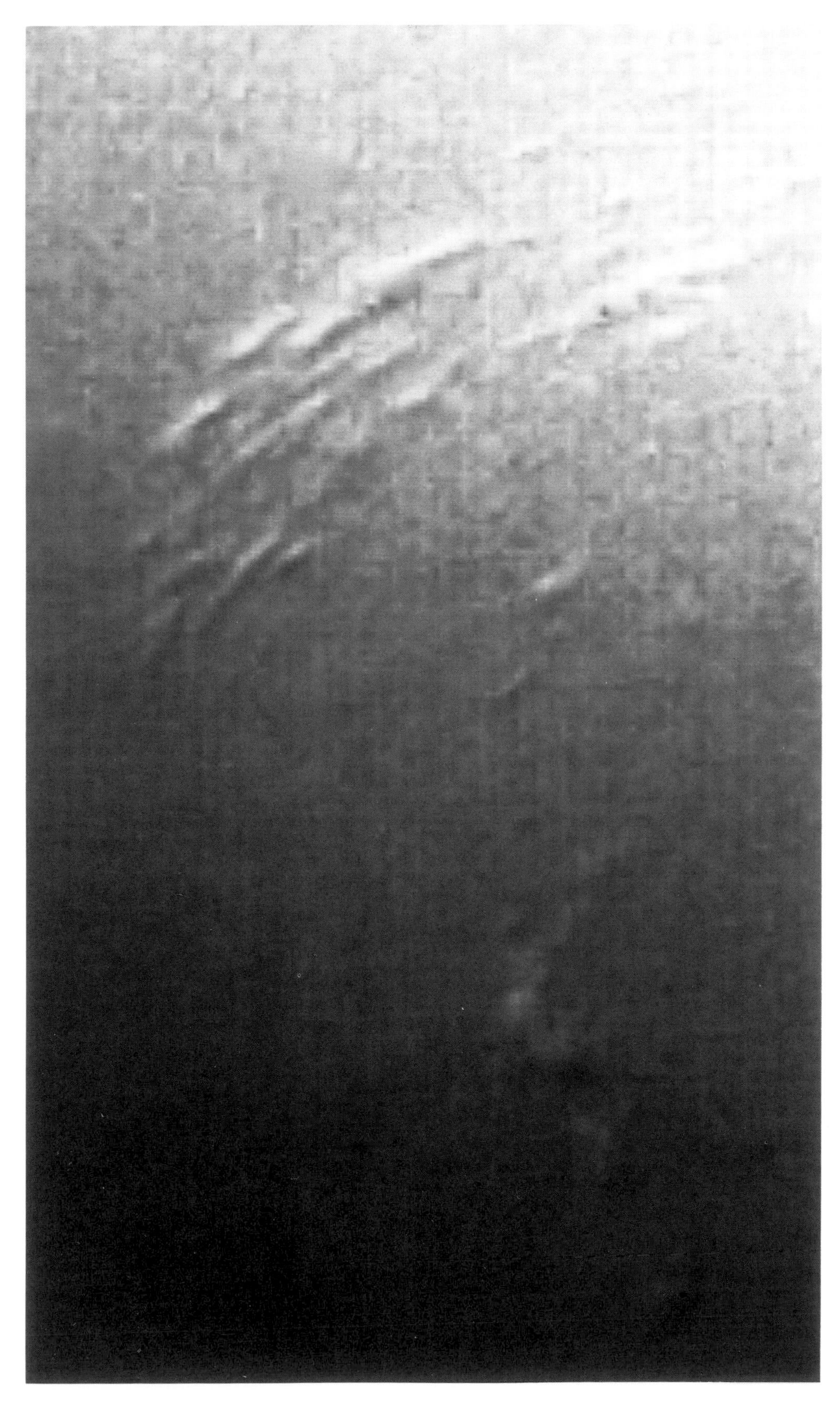

FIGURE 4.7 (A) This image of Neptune's south polar region shows cloud features the smallest of which is 28 miles (45 km) in diameter. The image also shows shadows cast on the lower cloud layers by the high clouds.

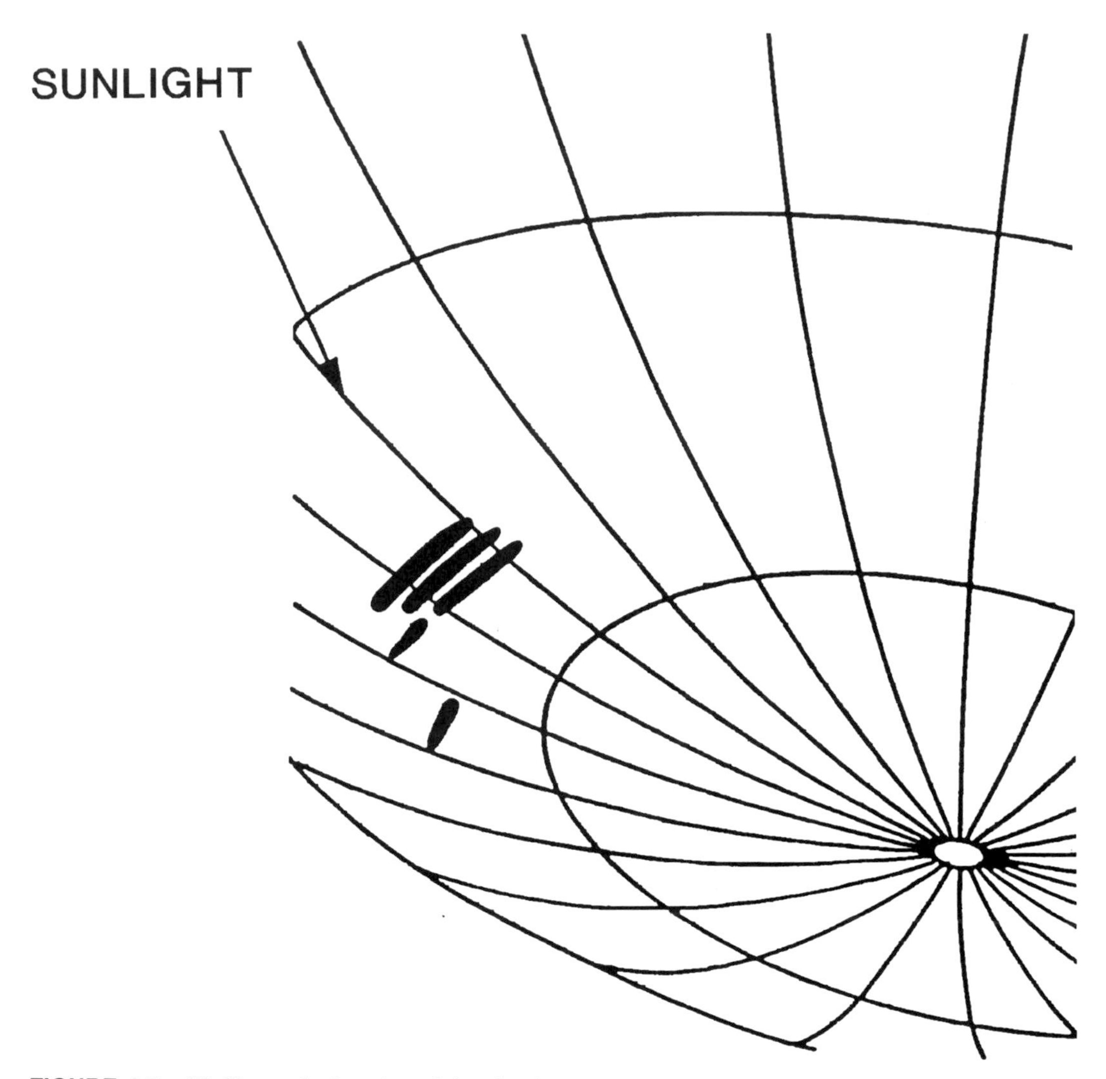

FIGURE 4.7 (B) Shows the location of the clouds at about 68 degrees south latitude. The discovery of cloud shadows on Neptune was the first observation of cloud shadows on any of the outer planets. (NASA/JPL)

tribution deep within the atmospheric shell. It seems that atmospheric convection results in cells analogous to Hadley cells (figure 4.13, p. 73) and in compressional heating. As gases rise in the atmosphere at mid-latitudes they cool and drift toward the equator or the poles at high altitudes and there they descend. As they move into denser regions, pressure is increased and the gases warm at the equatorial and the polar regions.

As expected, hydrogen is the main constituent of Neptune's atmosphere, and methane, at least in the upper atmosphere, is more abundant on Neptune than on Uranus. The high atmosphere's composition proved to be 85 percent hydrogen, 13 percent helium, and 2 percent methane. Occultations provided additional data about constituents of the deeper atmosphere where acetylene was identified and methane ice was detected. Also, there is a small amount of ammonia in the deep atmosphere, inferred from the absorption of radio waves during the occultation experiment. No evidence of ultraviolet absorbing aerosols was found in Neptune's stratosphere.

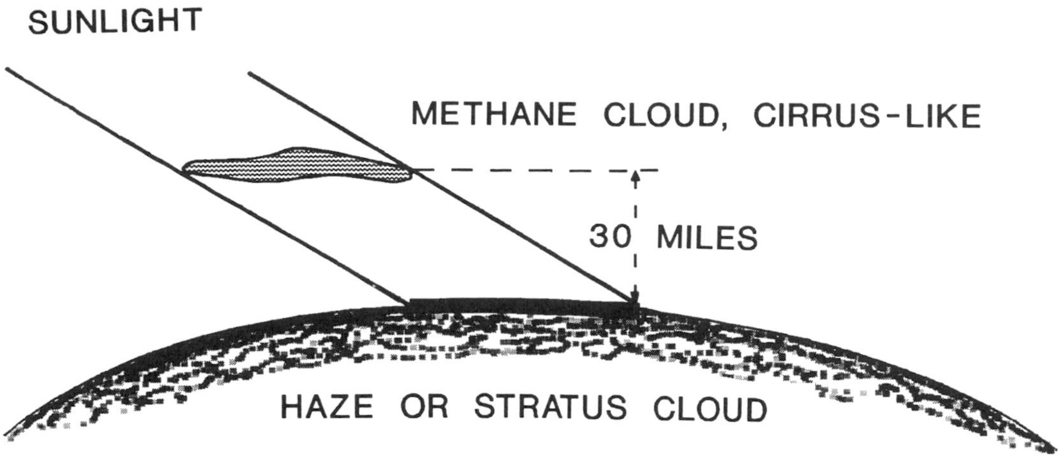

SUNLIGHT

METHANE CLOUD, CIRRUS-LIKE

30 MILES

HAZE OR STRATUS CLOUD

FIGURE 4.8 As shown in this sketch, the heights of the clouds above the main cloud deck is calculated by triangulation based on the altitude of the Sun and the angle its rays make with the surface of the cloud deck.

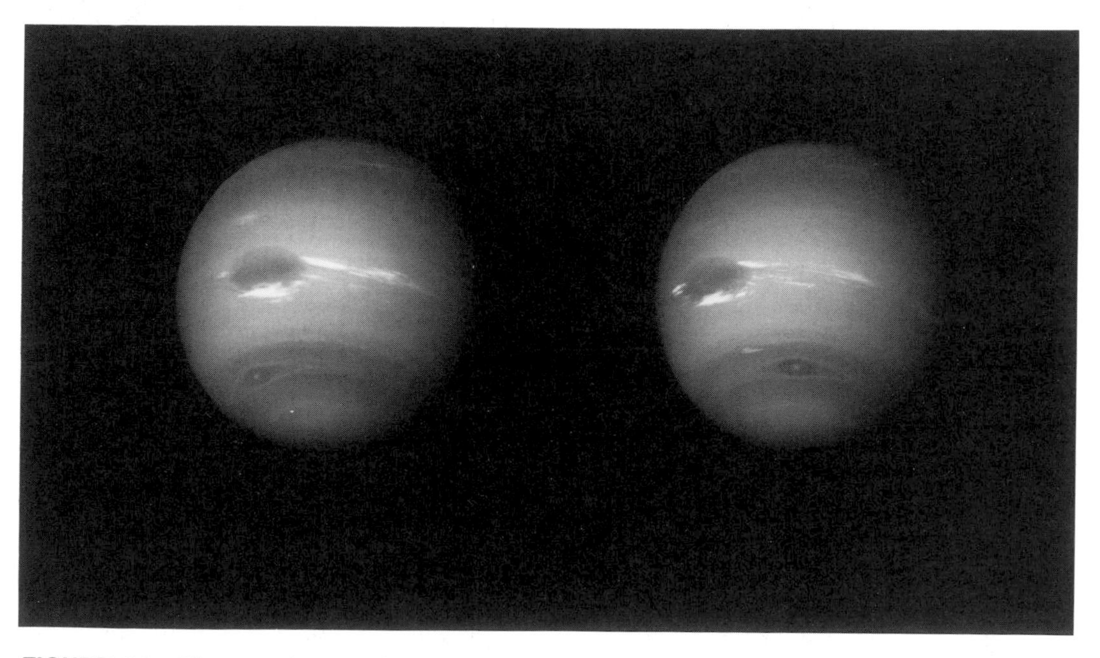

FIGURE 4.9 These two images of Neptune were taken 17.6 hours apart. In this period the Great Dark Spot at latitude 22 degrees has completed a little less than one rotation of Neptune while the smaller dark spot at 54 degrees south has completed a little more than one revolution. Light and dark bands circling Neptune indicate zonal motion around the planet. (NASA/JPL)

As *Voyager* closed in on the planet, the series of images it obtained (figure 4.14, p. 75) enabled scientists to track the cloud systems and determine wind speeds. Many were surprised to discover that, while the Great Dark Spot moved around Neptune in 18 hours, which was consistent with the period of rotation derived from ground-based observations, other regions relatively close by contained features that moved much faster. As mentioned earlier the bright "Scooter," for example, moved around Neptune

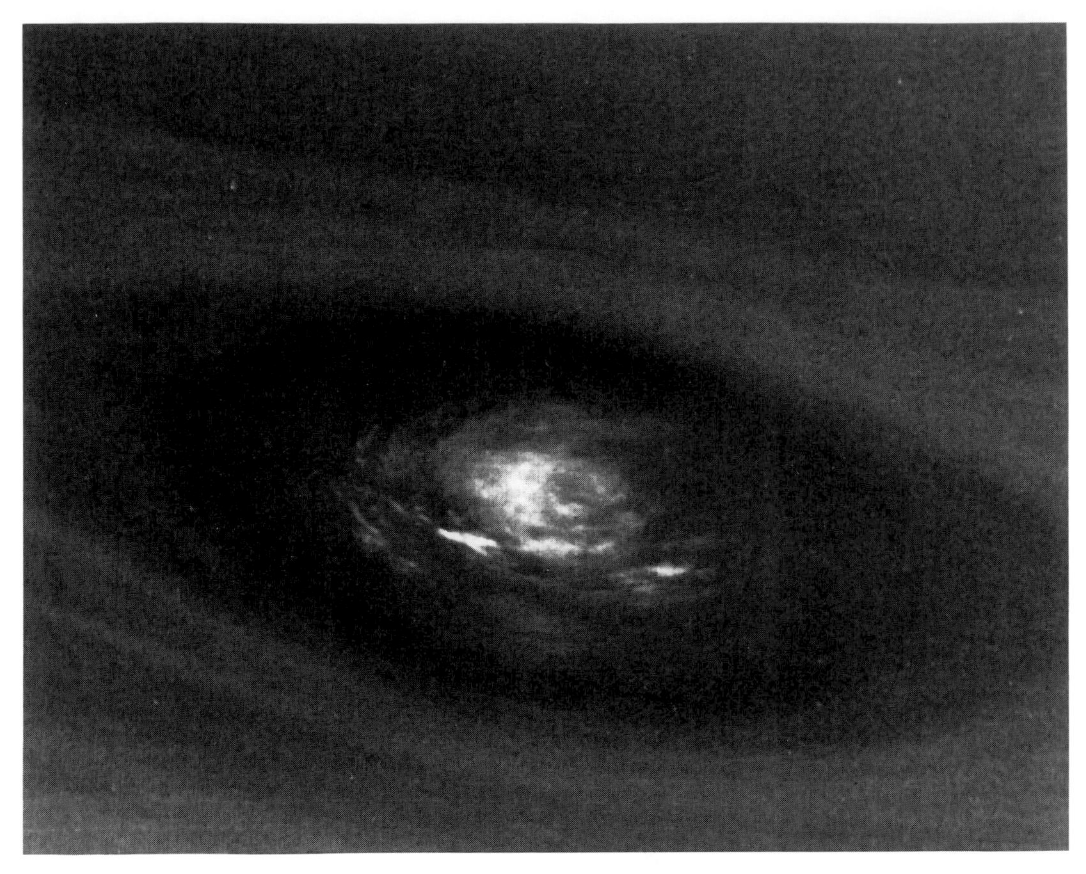

FIGURE 4.10 This close-up view of Neptune's small dark spot shows structures as small as 12 miles (20 km) across. Banding around the spot indicates the presence of strong winds, and the structure inside the spot suggests atmospheric upwelling and rotation of clouds around the center. A V-shaped structure near the right edge of the bright area suggests that the rotation is clockwise. (NASA/JPL)

in a period of only 16.1 hours. From the periodicity of Neptune's radio emissions scientists had determined that the bulk of the planet rotates in a period of 16 hours and 7 minutes. The wind velocities were accordingly calculated with respect to the basic period of rotation (figure 4.15, p. 76). Most of the winds on Neptune blow at up to 730 miles per hour (1180 kph), in a direction opposite to the direction of rotation of the planet. Thus, most of the atmosphere lags behind the bulk of the planet. However, some atmospheric features were observed moving at 45 miles per hour (72 kph) in the same direction as the rotation of the planet. The high easterly winds were at latitude 54 degrees south, and the top westerly winds were at latitude 22 degrees south. Near the Great Dark Spot there are winds blowing at the strongest encountered on the planet. Moving at up to 930 miles per hour (1500 kph), they cause severe wind shears near the spot. By comparison top winds on Jupiter blow at 335 miles per hour (540 kph), those on Saturn blow at 1120 miles per hour (1800 kph), and on Uranus at 450 miles per hour (720 kph).

A preliminary value for the hydrogen/helium ratio of the atmosphere used to derive temperature from the infrared spectrometer and the radio occultation data was 15:85 in terms of number density. This value will be refined later if found necessary.

The atmospheric chemistry is consistent with methane cycling from low to high altitudes. Solar ultraviolet radiation acts on methane high in Neptune's atmosphere, converting it into hydrocarbons such as ethane and acetylene and smog-like haze particles of more complex polymers. These heavier gases and particles freeze into various ice crystals to produce clouds. Some sink to the lower atmosphere, slowly falling into warmer regions where they evaporate back into gases. Deep within the atmosphere, where the temperature is about 1000 K (1300° F), these hydrocarbon gases mix with hydrogen; at the higher temperatures and pressures they are converted back into methane. The methane forms buoyant convective clouds that rise high into the atmosphere to form plumes at two latitude bands and thereby return methane gas back to where it can again be acted upon by solar ultraviolet radiation. There is no net loss of methane gas from the atmosphere.

The methane ice clouds are in the troposphere at the 1 bar level. About 30 miles (48 km) below them are thick dense clouds of hydrogen sulfide ice at the 3 bar level.

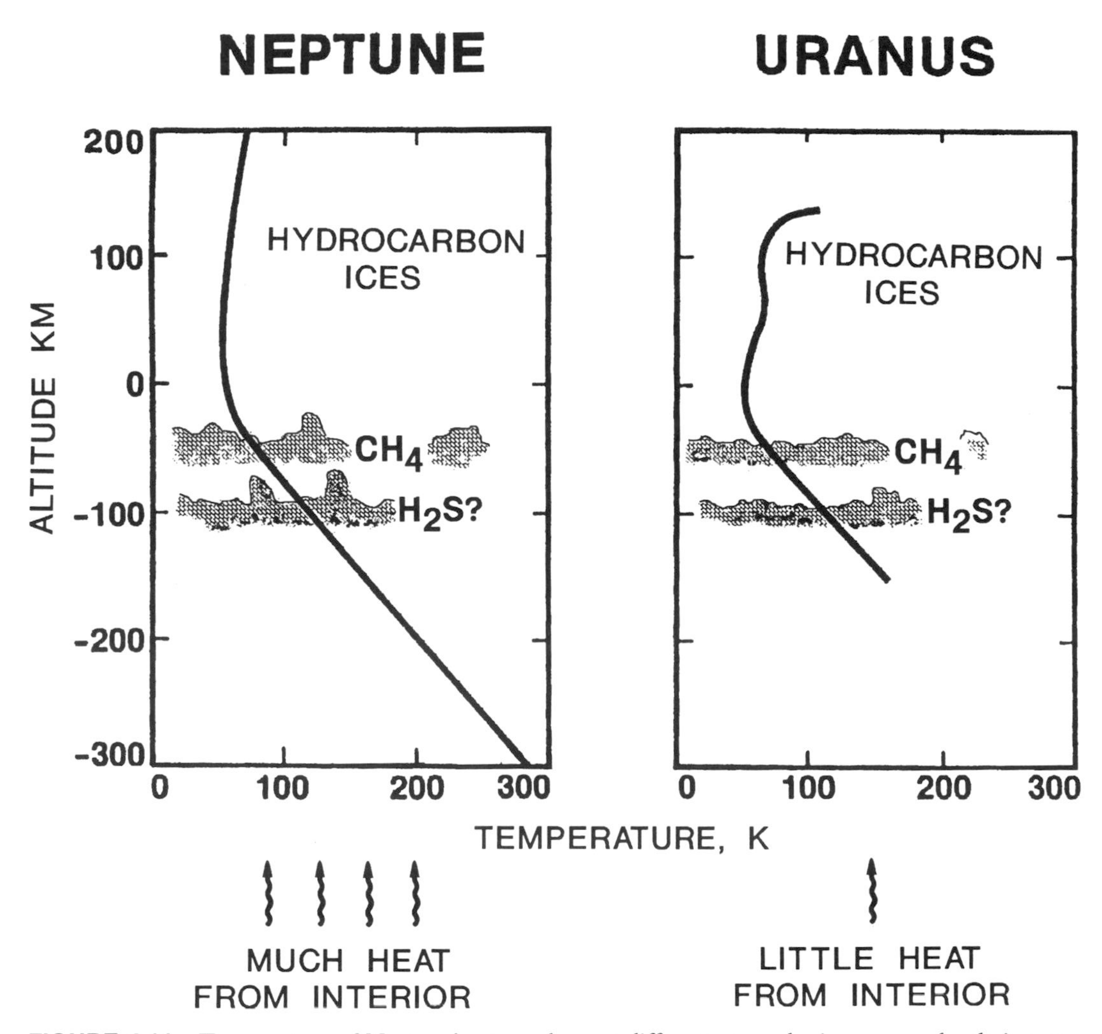

FIGURE 4.11 Temperature of Neptune's atmosphere at different atmospheric pressure levels is compared with that in Uranus's atmosphere. Cloud levels are also indicated for the main cloud decks. (After NASA/JPL)

FIGURE 4.12 An infrared map of Neptune obtained by the IRIS experiment shows cooler bands (lighter areas) at mid-latitudes compared with the equator and the poles. (After JPL)

Hydrocarbon smog of low optical density extends over 90 miles (150 km) of the lower stratosphere. Filters on the optical instruments provided information to locate the levels of the clouds in addition to the offset shadow calculations. Some ultraviolet light was also scattered to lighten the shadows cast by the high clouds (figure 4.16, p. 76), and from the amount of the scattered light information could be derived about the nature of the particles scattering the ultraviolet light. Polarimetry data indicate that at Neptune, as at Uranus, hazes developed in the stratosphere by incoming radiation and charged particles are optically very thin.

The light companion to the Great Dark Spot, associated with its southern edge, is 30 miles (50 km) above the methane clouds. The "Scooter" is low in the atmosphere because it is hidden by methane absorption. It appears to be a plume rising from the hydrogen sulfide cloud layer.

The cloud systems and temperature curves for the four giant planets are compared in figure 4.17, p. 77. The ultraviolet spectrometer measurements showed that in Neptune's upper atmosphere at a level where the pressure is 10^{-12} bar, the temperature is 400 K (260° F) which is almost half that at the same level on Uranus. Some of the differences between the atmospheres of Uranus and Neptune most probably arise because Neptune's atmosphere receives heat from the planet's interior, while Uranus' atmosphere receives very little heat from within the planet.

In the high atmosphere there is an extensive ionosphere with multiple layers between

620 and 2500 miles (1000 and 4000 km) above the level at which the atmospheric pressure is 1 bar.

The first observation confirming that Neptune has a magnetic field was when periodic radio emissions from the planet were detected by the plasma wave instrument some 30 days before closest approach. The instrument also detected plasma waves produced by the bow shock on August 24, 1989. Estimates were made (table 4.1) on where the bow shock would be crossed, assuming different values for the magnetic field at the visible "surface" of the planet: the cloud tops.

However, as the spacecraft continued in, it was expected that a bow shock would appear in the data from other instruments; but this evidence did not materialize. Nor

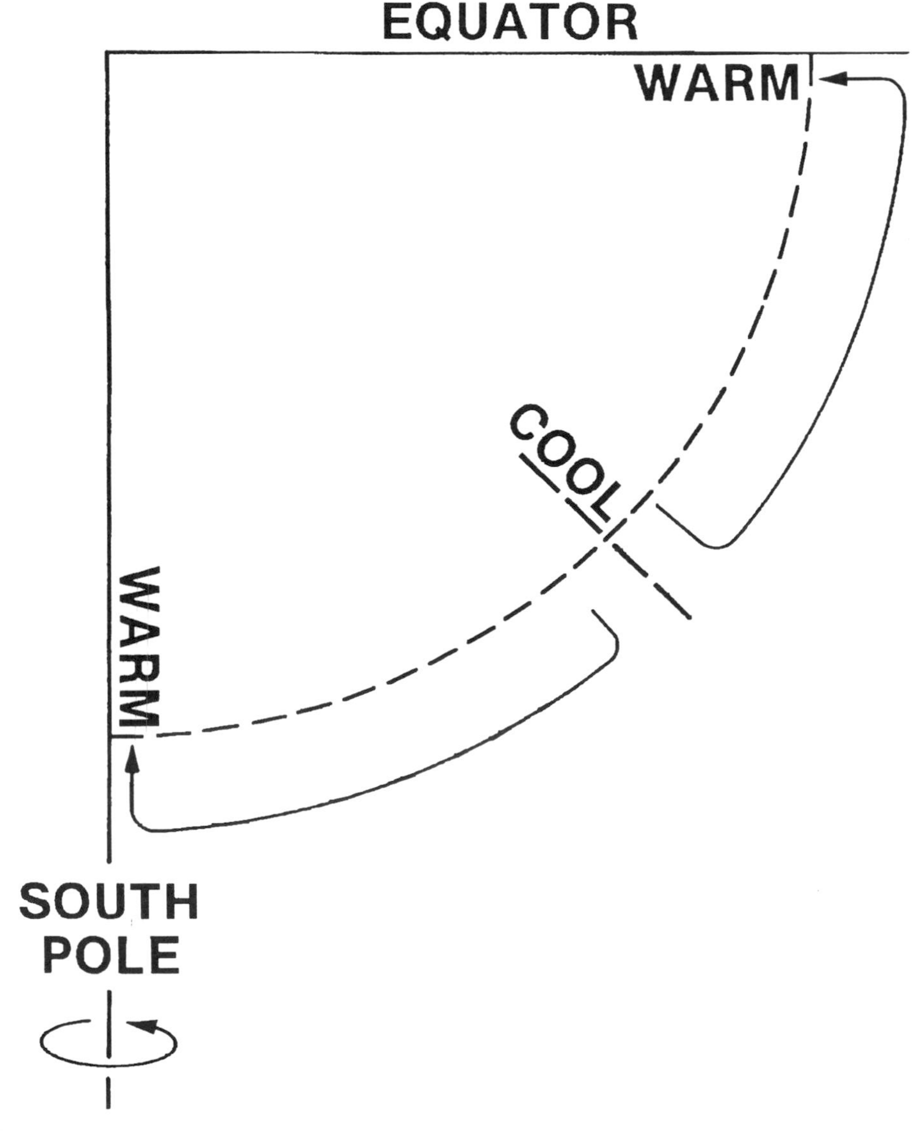

FIGURE 4.13 Circulation suggested for Neptune's atmosphere to account for the temperature distribution. Atmosphere ascends in temperate regions and cools as it expands. The atmospheric mass flows toward the equator and the poles where it descends and is heated by compression.

TABLE 4.1 Predicted Crossings of the Bow Shock and Magnetopause

Surface Field (Gauss)	*Bow Shock Distance**	*Magnetopause Distance**
4.0	78	60
1.0	49	38
0.31	33	26
0.2	29	22

*In planetary radii

had *Voyager* observed any of the characteristic upstream ion events or energetic charge-exchange neutrals to show the presence of a magnetosphere and its interaction with the solar wind. Such events had been detected at Jupiter and Saturn hundreds of planetary radii ahead of the bow shock. Ion events were also detected well ahead of the bow shock at Uranus. The reason for this delay in detecting upstream events at Neptune that would indicate the presence of a magnetosphere, and the delay in passing through a shock would be revealed later when an unexpected tilt of the magnetic field to the rotational axis of the planet was discovered.

The bow shock is where the solar wind encounters a magnetic field of sufficient intensity to slow the wind, heat it, and deflect it around the planet (figure 4.18, p. 77). Inside the bow shock is the magnetosphere of the planet in which the planetary magnetic field controls the motions of charged particles such as ions and electrons. Between the bow shock and this magnetosphere is a region referred to as the magnetopause, which separates the shocked solar wind particles and the particles confined within the magnetic field of the planet comprising the magnetosphere. At the magnetopause the magnetic field of the planet holds off the solar wind the particles of which are forced to flow around the planet.

At 9:20 A.M. on August 24 the bow shock had not been detected by *Voyager*'s instruments. Edward C. Stone, the project scientist, commented at a press conference that "We're seeing some unusual conditions in the solar wind that we don't understand yet." At this time the solar wind was flowing at a speed of 901,000 miles per hour (1,450,000 kph). Its temperature was 6300 K, and its density was 0.0045 protons per cubic centimeter. Electrons were not energetic enough to be detected in the solar wind by the plasma science instrument (PLS). Later that same morning telemetry data arrived at 11:46 A.M. PDT, after its four-hour journey from Neptune, to show from the data that *Voyager* actually crossed the bow shock at 7:40 A.M. PDT (at the spacecraft), less than 12 hours before its closest approach to the planet. There was only one very regular bow shock crossing on this inward leg of the trajectory, compared with two or three for the other large planets. This probably resulted from the flow of the solar wind being relatively calmer and steadier at Neptune's great distance from the Sun.

As data continued to arrive the bow shock crossing was confirmed by all three instruments: the magnetometer, the plasma wave science, and the plasma science experiments. The magnetometer showed an abrupt change in the magnetic field at 11:42 A.M. PDT; the plasma wave science showed evidence of the shock at 11:32 A.M. PDT, and the plasma experiment showed changes in ions and electrons at 11:30 A.M. PDT. Ions and electrons were heated to a much higher temperature by passage through

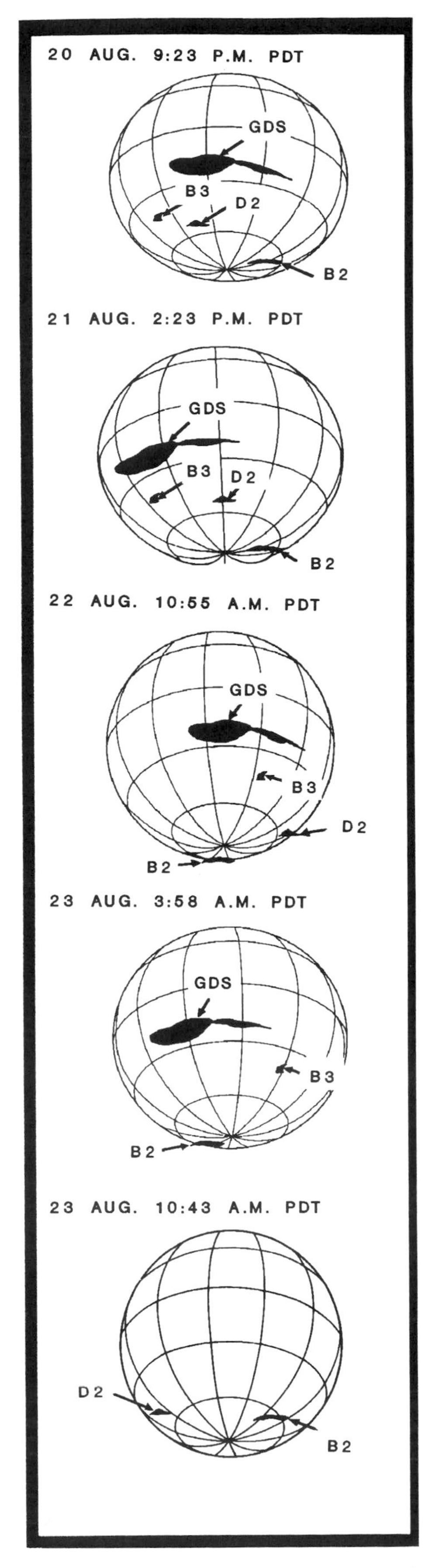

FIGURE 4.14 A series of images showing the motion of markings relative to one another provided information to determine the direction and speed of winds at various latitudes on the planet. These drawings show how the features changed at various rotations of the planet as captured on images at the dates and times shown. (JPL)

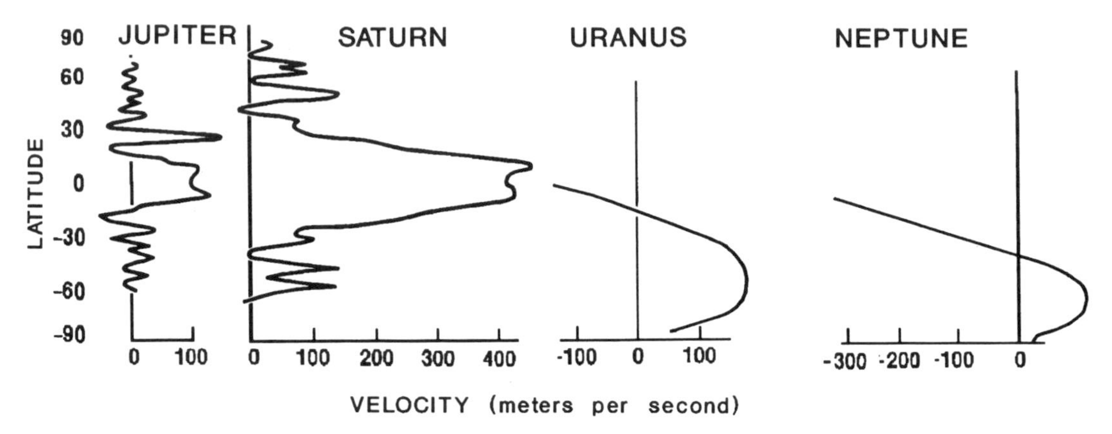

FIGURE 4.15 Plots of wind speeds for the four giant planets; the greatest speeds are encountered on Saturn and Neptune. Winds on Jupiter and Saturn favor one direction so that equatorial regions rotate faster than the rotation of the whole planet. Winds on Neptune and Uranus, like the Earth, favor the opposite direction with equatorial air masses rotating around the planet slower than the planet's rotation. Winds on Neptune when extrapolated from the observations would appear to be the highest of all planets. (After JPL)

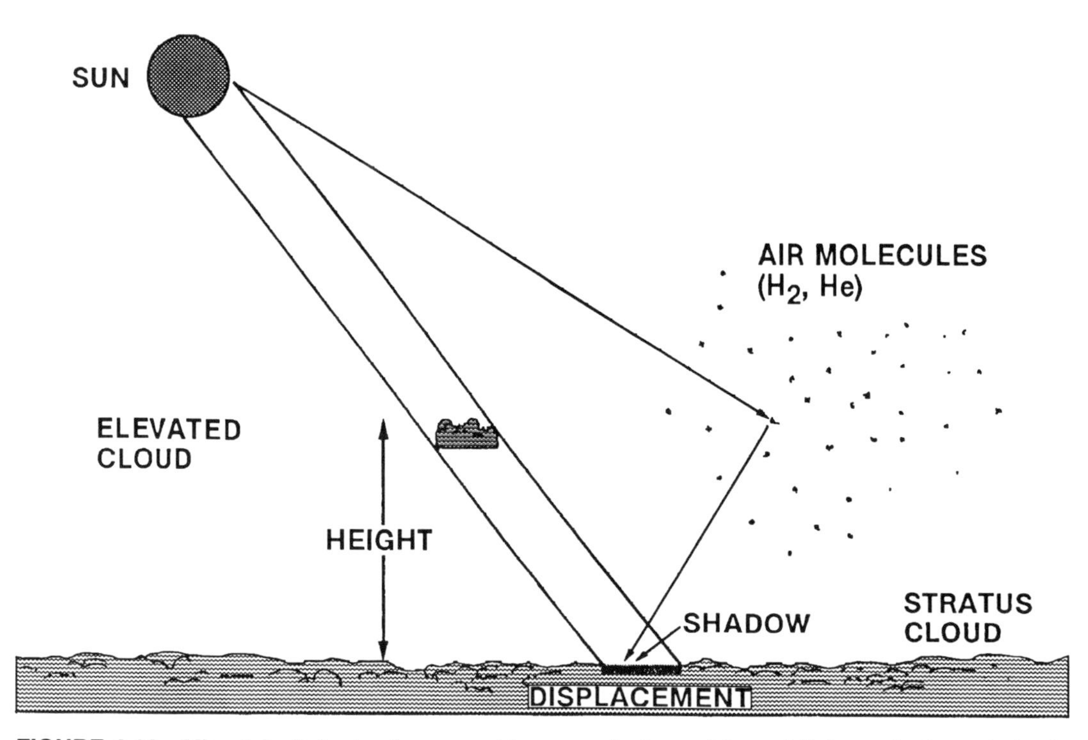

FIGURE 4.16 Ultraviolet light that is scattered by atmospheric particles and lightens shadows of clouds provides information about the amount of haze in the atmosphere above the main cloud deck.

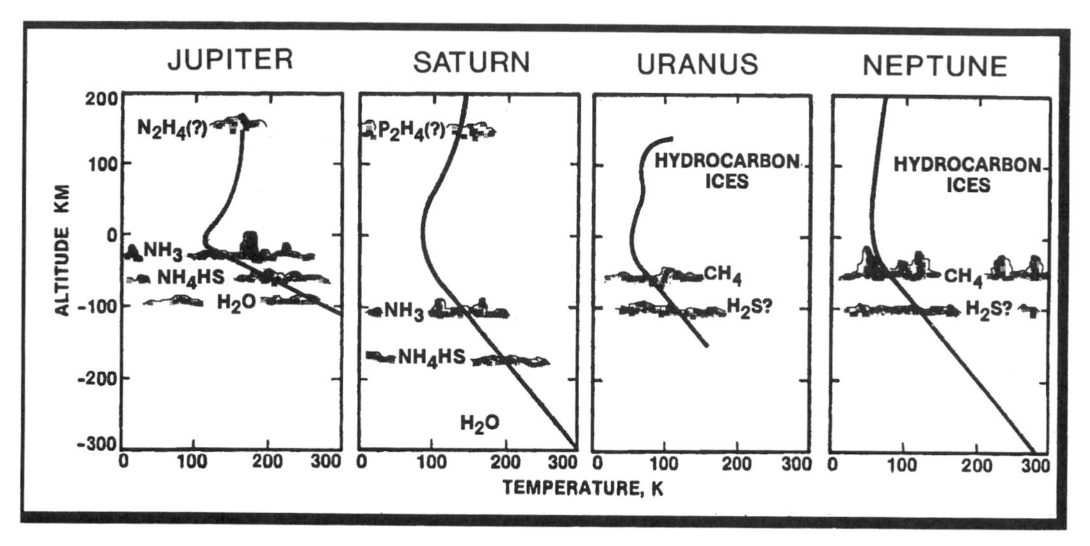

FIGURE 4.17 The temperatures of the atmospheres of the four large planets at various pressure levels are compared in this drawing. (After JPL)

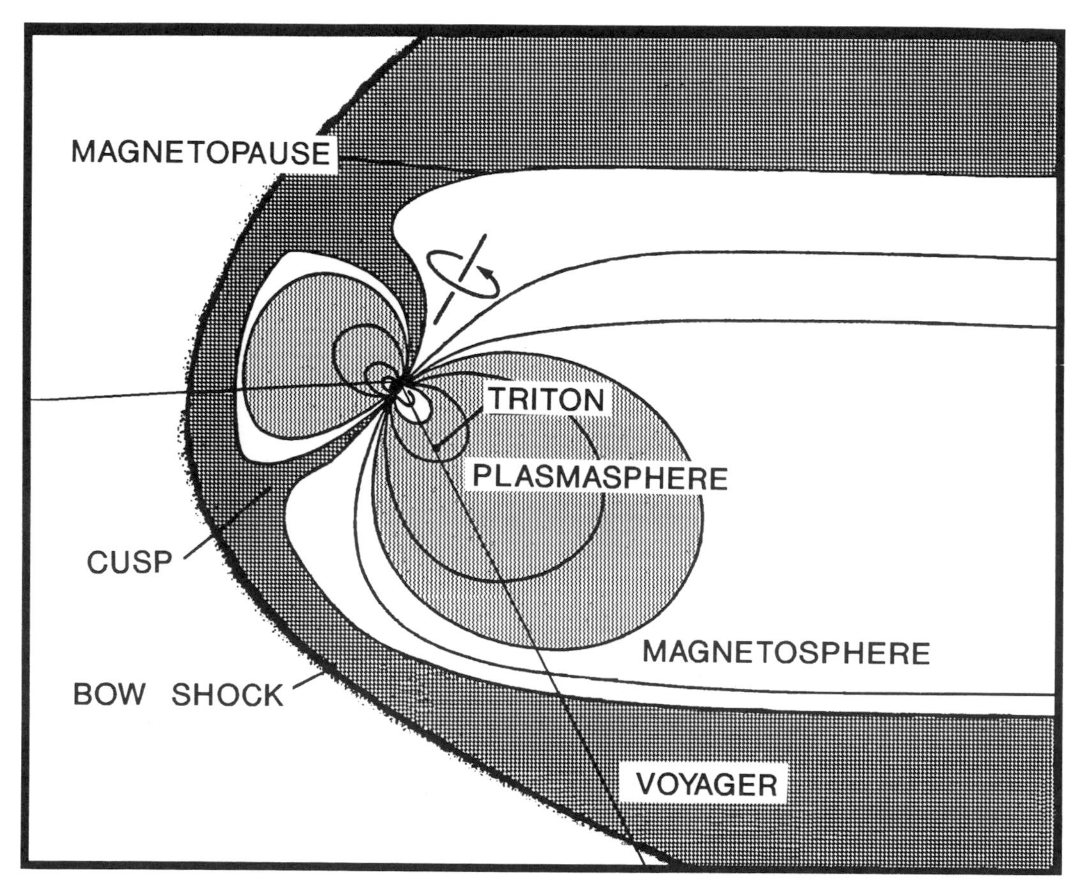

FIGURE 4.18 The magnetic environment expected at Neptune is shown in this figure together with the path of *Voyager* through the system and the orbit of Triton. A bow shock was expected followed by evidence of the magnetopause with the spacecraft later passing through the cusp associated with the north magnetic pole.

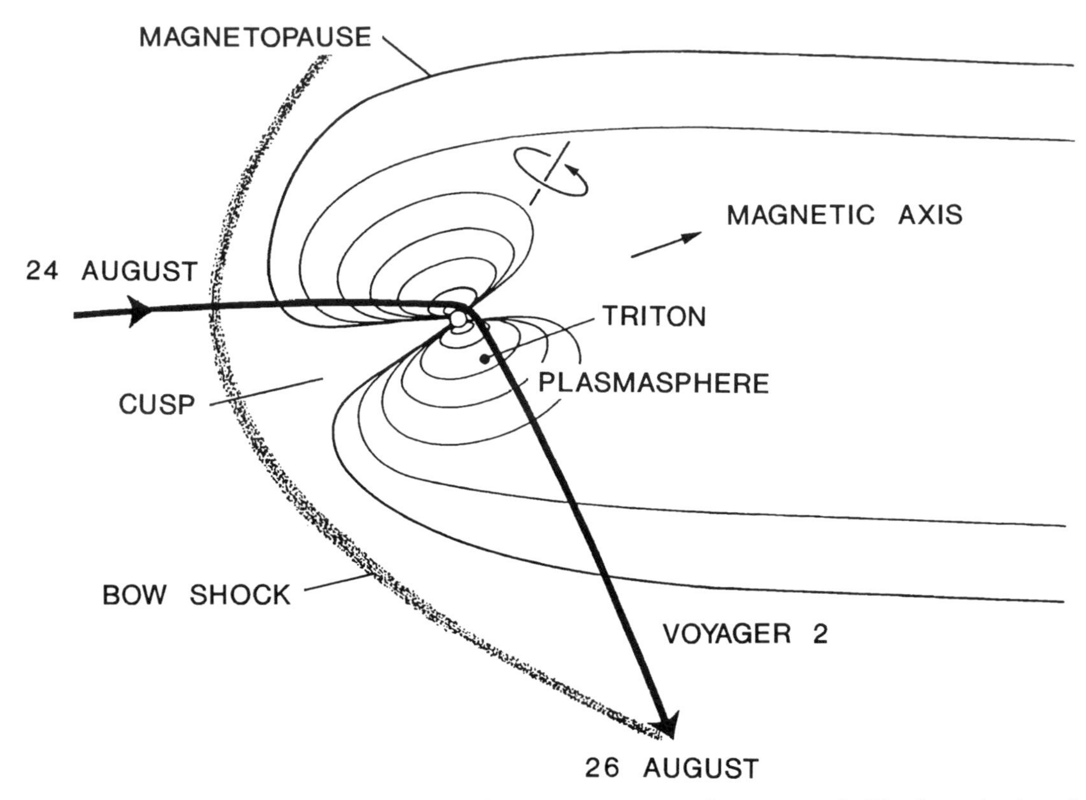

FIGURE 4.19 Surprisingly the magnetosphere was not oriented as expected. The bow shock and magnetopause were not encountered as expected and the spacecraft approached the planet very close to one of the cusps of the magnetosphere. This was because of an unexpected orientation of the planet's magnetic axis to its axis of rotation. (JPL)

the shock so that the electrons as well as the protons could be detected by the PLS experiment. The spacecraft remained within the magnetosphere for about 38 hours before crossing the shock again and emerging back into the solar wind on August 26 (figure 4.19). *Voyager* then experienced several bow shock crossings through August 28 as it traveled away from the planet.

The crossing through the magnetopause into the magnetosphere was not so clearly defined as the crossing of the bow shock. This was a result of the spacecraft entering the magnetosphere near a magnetic pole. As the spacecraft flew through the magnetosphere, the plasma and low-energy charged-particle instruments collected data about the magnetic field. It was soon apparent that the field was tipped more than 30 degrees from the planet's axis of rotation. Also the field had to be offset from the center of the planet. "We had expected to find a magnetic dipole roughly aligned with the rotation axis," John Belcher, principal investigator for the plasma science instrument, told the media at a press conference. "We also expected to see plasma concentrated toward the equator, plasma in the magnetic cusp near the (geographic) pole, and a sharp change as *Voyager* crossed the magnetopause." Instead *Voyager 2* found a tipped field, experienced an extended crossing of the magnetopause, and detected a plasma sheet that was not in the planet's equatorial plane. It became clear that the spacecraft had entered the magnetosphere close to a cusp.

The magnetic field results of Belcher were confirmed by the low-energy charged-particle experiment for which Tom Krimigis was principal investigator. This instrument measured high-temperature, low-density electrons in the magnetic field of the planet. While the temperature of the electrons was about 1.3 billion degrees, their concentration was low. The high temperatures of particles within the plasma do not affect the temperature of the spacecraft because, although incredibly hot, the highly rarefied plasma has too little mass per unit volume to transfer significant quantities of heat to the spacecraft.

Later during the encounter scientists gathered and analyzed more data to show that Neptune's magnetic field is tilted 47 degrees from the planet's axis of rotation, and it is also offset more than half the radius of Neptune from the center of the planet; by 0.55 radii, which is approximately 8,460 miles (13,620 km), (figure 4.20). At a press conference, Norman F. Ness, principal investigator for the magnetic field experiment, explained how the tilted offset field produced great variations in the magnetic field strength at the cloud tops (figure 4.21). Near the magnetic pole in the southern hemisphere of Neptune the field is strongest and amounts to 1.2 gauss whereas at the

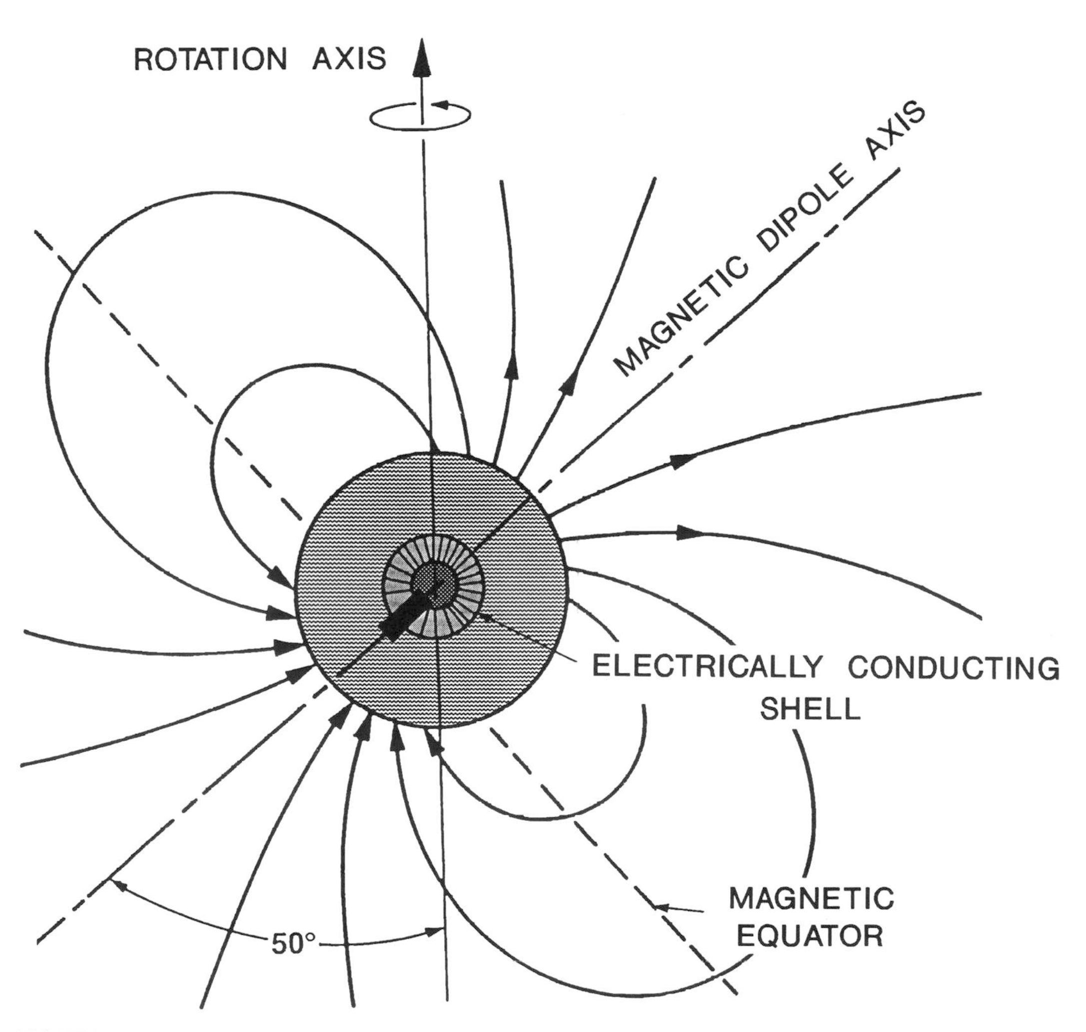

FIGURE 4.20 The orientation of the planetary dipole is shown here relative to the rotation axis. The dipole is also displaced outward from the center of the planet as shown in the drawing.

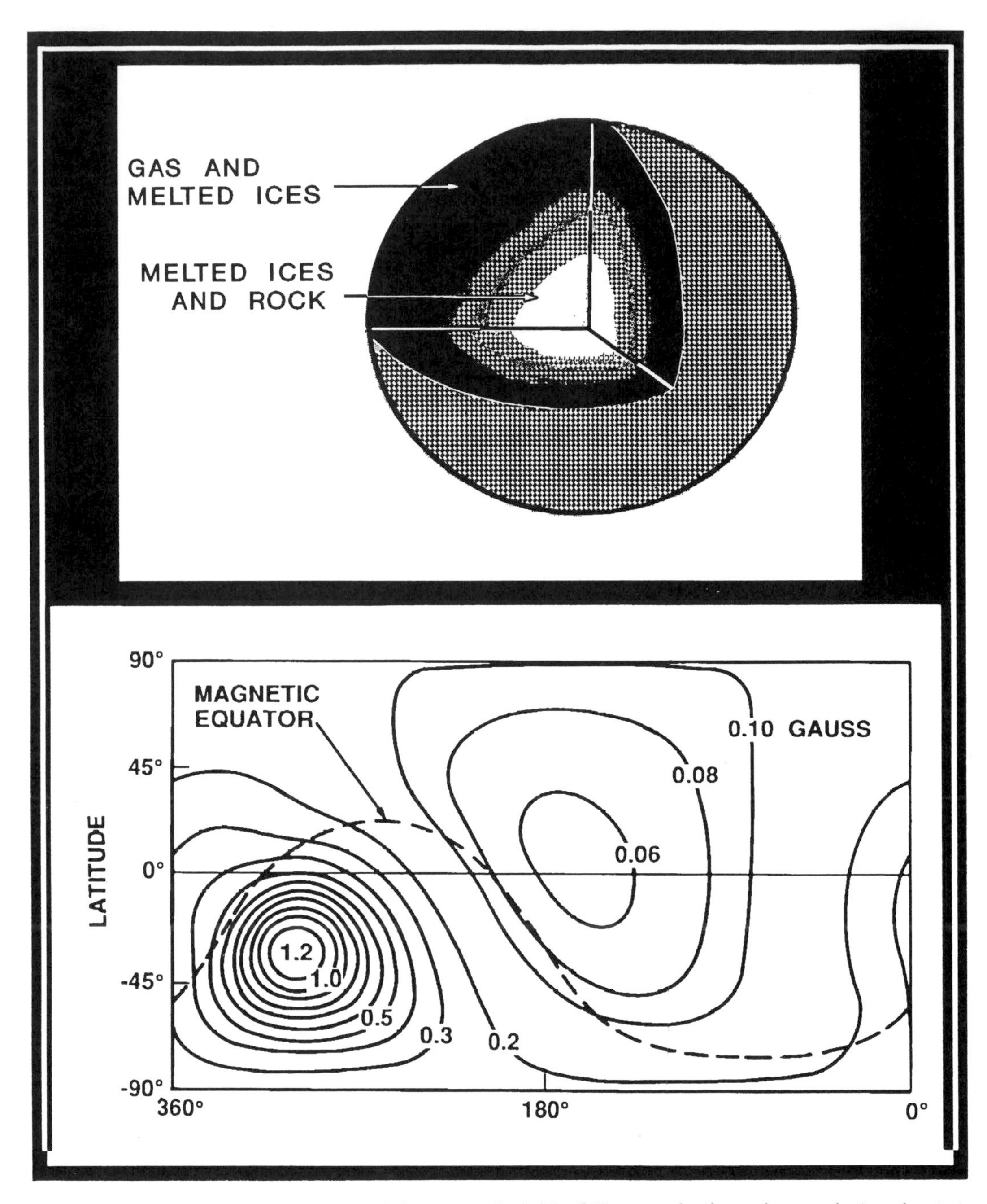

FIGURE 4.21 (A) The orientation of the magnetic field of Neptune leads to the conclusion that it is generated in a shell close to the surface rather than in the core of the planet. (B) The strengths of the fields at the main cloud layer and the location of the magnetic equator are shown on this drawing. (JPL)

magnetic pole in the northern hemisphere of the planet the field is much weaker (0.06 gauss) with the magnetic field lines further apart. The magnetic equator is a complicated curve on a Mercator-type plot of the planet. At the magnetic equator Neptune's field's strength is 0.133 gauss. Earth's field, by comparison, is 0.31 gauss at the terrestrial equator and Earth's field is of opposite polarity to Neptune's field.

Neptune's field is not a simple dipole; it is complicated with evidence of quadrupole and octupole components and possibly local anomalies. The complexities of the field suggest that there are dynamo electric currents close to the outer regions of the planet. In all the large outer planets the quadrupole and octupole components are significant because the regions where the magnetic field is generated are not deep in the planetary core but are in a shell or shells closer to the visible surface. How close is not yet known. In Jupiter and Saturn the internal pressures are high enough to compress hydrogen into liquid metallic hydrogen in which dynamo currents can flow deeper than in Uranus and Neptune, where another conductive fluid provides the medium for the dynamo currents closer to the surface.

Magnetospheric scientists compared Neptune's field with that of Uranus, which is tilted 59 degrees from the rotation axis and with a center offset by 0.3 planetary radii. After *Voyager 2*'s encounter with Uranus some scientists speculated that *Voyager* might have witnessed an actual reversing of the direction of Uranus's field during the flyby. Earth's field has reversed several times during the planet's history, as evidenced by analysis of rocks. Similar speculations followed the discovery of a highly offset field at Neptune. However, other scientists found it difficult to believe such coincidences. Instead they suggest that the tilt reflects flows in the interiors of the planets that are unrelated either to the tilt of Uranus's rotational axis or to field reversals at either planet. One thing appears certain since the magnetic field of Neptune has such a large tilt. The speculation that the tilted magnetic field of Uranus resulted from witnessing a field reversal is unfounded. It would be most unlikely to encounter fields reversing at two planets about the same time.

When observed from space around the planet, the magnetic field of Neptune goes through dramatic changes as the planet rotates and the magnetic axis is swung around in the solar wind.

Because the spacecraft's trajectory brought it in toward the magnetic pole, *Voyager* provided the first observation of this type of interaction of the magnetosphere with a planet's magnetic field.

Fortuitously the timing of *Voyager*'s encounter with Neptune proved to be such that the south magnetic pole was pointed generally toward the Sun, so that the solar wind was blowing into the northern cusp of the magnetosphere. Because the solar wind more easily penetrates toward the magnetic pole, the bow shock was close to the planet, and this explained why *Voyager* seemed so long in crossing the bow shock. While completely unexpected this orientation of the magnetic field provided important science dividends. *Voyager* experimenters were given a unique opportunity to observe conditions in the cusp region; the first time for such a gigantic magnetic field.

The 47-degree tilt of the magnetic field of Neptune is somewhat less than the tilt of Uranus' field, which is 58.6 degrees, but different from Jupiter, Saturn, and Earth. The dipole offset from Neptune's center by 0.55 planetary radii is the largest offset so far measured in the Solar System. The polarity of Neptune's field is the same as those of Jupiter and Saturn, but opposite to that of the Earth. Comparisons of the magnetic fields and their orientations for the four giant planets and for the Earth and Mercury are shown in table 4.2.

The theory of planetary magnetism postulates that to have a magnetic field a planet should possess an interior region that is liquid and electrically conducting (figure 4.21). Also the planet must provide a source of energy to keep this region moving to generate

TABLE 4.2 Comparisons of Planetary Magnetic Fields

Planet	Magnetic Field (Gauss) Min.	Dipole Equator	Max.	Dipole Tilt & Sense	Offset Radii	Magnetopause Distance* Radii
MERCURY	0.0033	0.0033	0.0066	+14	0.05	1.4
EARTH	0.24	0.31	0.68	+11.7	0.07	10.5
JUPITER	3.2	4.28	14.3	−9.6	0.14	65
SATURN	0.18	0.215	0.84	−0.0	0.04	20
URANUS	0.08	0.228	0.96	−58.6	0.3	18
NEPTUNE	0.06	0.133	1.2	−47	0.55	25

*Varies with pressure of the solar wind

a field. According to this theory, the character of a planet's field provides important information about conditions in the interior of a planet. To account for Neptune's field the dynamo electric currents produced within the planet must be relatively closer to the surface than for Earth, Jupiter, or Saturn.

The interior of Neptune appears to be as expected; a central core of ionized water, melted ices, and rock, with rocks concentrated toward the center, an outer, 6000-mile (10,000 km) thick shell of gas and melted ices, and a deep atmosphere with a dense cloud layer. Even at the center of the planet there is probably no solid planetary type core but rather a hot fluidized mixture of water and rocky materials under enormous pressure and at a very high temperature. Also, the various shells are unlikely to have sharp and distinct boundaries but instead merge gradually one into another. The skewed magnetic field, as with Uranus, probably originates from the convection of electrically conducting material in a shell region instead of within the core as is believed to occur at Earth.

An important piece of information derived from the observations of Neptune's magnetic field was a confirmation of the rotation rate of the planet. Observers on Earth could not determine the precise length of the day on Neptune since observations of the motions of large cloud features across the face of the planet have several variables that are difficult to isolate. Cloud motions are affected by winds that in turn vary with latitude. The best estimates of a rotation period derived from telescopic observations of Neptune was one of 18 hours, but this was an approximate period only. However, the rotation rate of the interior of a planet can be derived from periodic radio waves generated by the magnetic field. From this, as mentioned earlier, the radio scientists determined that Neptune rotates in a period of 16 hours and 7 minutes.

Voyager detected auroras similar to the northern and southern lights we observe on Earth. Auroras occur when energetic particles strike the atmosphere as they spiral down magnetic field lines toward the magnetic poles (figure 4.22). The auroras concentrate in circles around these magnetic poles and are thus most frequently seen at high latitudes in Earth's Northern and Southern Hemispheres. The precipitating particles detected by the low-energy charged-particle (LECP) instrument at Neptune, mainly protons and electrons but with some heavier ions, cause the atmosphere to glow and emit radio waves. This was the first direct detection of auroral zone particles impacting the atmosphere of a planet other than the Earth. The sources of the particles—protons, ionized hydrogen molecules, and ionized helium—detected in Neptune's magnetos-

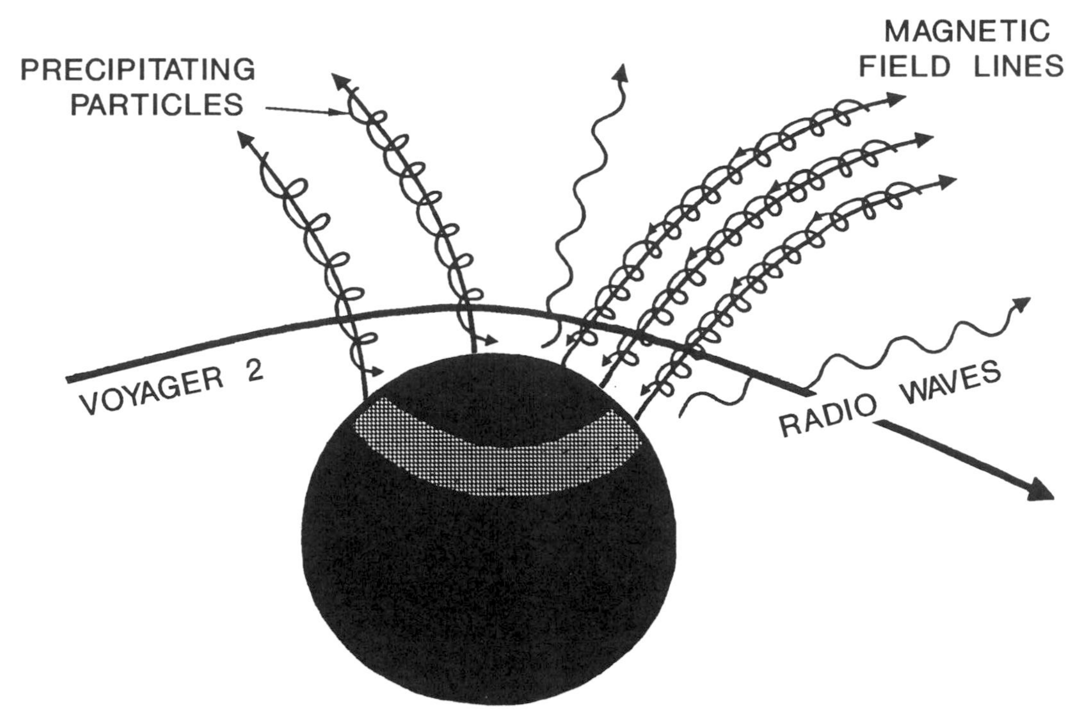

FIGURE 4.22 The Low Energy Charged Particle (LECP) experiment detected protons, electrons, and heavier ions precipitating into the atmosphere of Neptune to produce radio waves and aurorae.

phere could be the solar wind, the planet's ionosphere, and the satellites. A comparison of the ratios of the various particles including helium show an abundance of protons, secondly ionized hydrogen molecules, and thirdly helium. The measured ratios of numbers of particles were 1400:1:0.1. The hydrogen/helium ratio of these particles is not that of the solar wind, so it is concluded that the major source of these magnetospheric particles is Neptune's ionosphere. This is similar to conditions at Uranus but is different from Jupiter and Saturn, where the magnetospheric particles originate from both the ionosphere and the satellites.

Precipitation of particles from the magnetosphere was confirmed by analysis of radio waves detected by the plasma wave experiment. A particular type of radio signals called a "whistler" is generated by lightning flashes and by electrons in radiation belts. Whistlers have been observed at Earth, Jupiter, Saturn, and Uranus. They were also detected at Neptune but there was no confirmation of lightning at Neptune. In the radiation belts electrons spiral along lines of magnetic force and generate radio waves. If these signals travel back along the field lines, following electrons are affected and oscillations are generated that produce a characteristic hiss at a radio receiver. The effect of radio waves bouncing back and forth in the feedback process which produces the radio oscillations is to cause the electrons to precipitate from the radiation belts and fall toward the magnetic poles. These whistler mode radio noises were observed before and after closest approach of *Voyager* to Neptune, confirming that electrons were being lost to the radiation belts and precipitating into Neptune's atmosphere to cause auroras.

The auroras developed around the magnetic poles by particles precipitating from the magnetosphere, as detected by the LECP, may not necessarily be the same as those

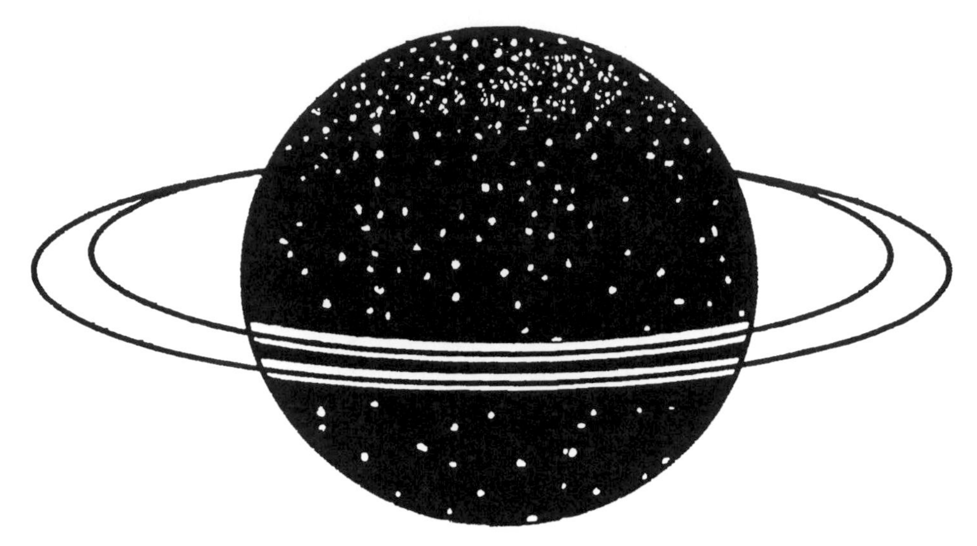

FIGURE 4.23 The Ultraviolet Spectrometer (UVS) data also contained evidence of aurorae but these were not concentrated in the magnetic polar regions. Instead they appeared to be widely distributed over the planet, unlike those observed by this instrument at Jupiter, Saturn, and Uranus.

TABLE 4.3 Auroral Power at Earth Compared with Outer Planets

Planet	*Auroral Power in Watts*
Earth	10^{11}
Jupiter	10^{14}
Saturn	10^{11}
Uranus	4×10^{10}
Neptune	10^{6}
Triton	10^{9}

observed by the ultraviolet spectrometer. These latter are more generally distributed over wide regions, rather than just close to the magnetic poles (figure 4.23). Arising from Neptune's complicated magnetic field they are distributed differently from those observed on Jupiter, Saturn, and Uranus, and the ultraviolet spectra of the Neptunian auroras differ from those of the other planets. The power of auroras at Neptune is thousands of times less than terrestrial auroras. Table 4.3 compares the power of auroras observed at the various planets and at Triton.

As the planet rotates, the magnetic field rotates with it so that in any given location in the region surrounding the planet the field varies greatly with time. Both the rings and the satellites continually experience a wide range of magnetic latitude changes. Also as the field rotates with respect to the Sun, the interaction with the solar wind varies enormously from the field being pole-on to the incoming solar wind to being a more normal attitude like that of Earth's field. The magnetotail also changes during each rotation of the planet and its field, varying from a normal magnetotail sheet when the magnetic equator aligns with the flow of the solar wind, to a cylindrically-shaped plasma sheet when the pole of the planet's magnetic field points toward the Sun (figure 4.24).

The plasma environment within the magnetosphere of the planet was explored by

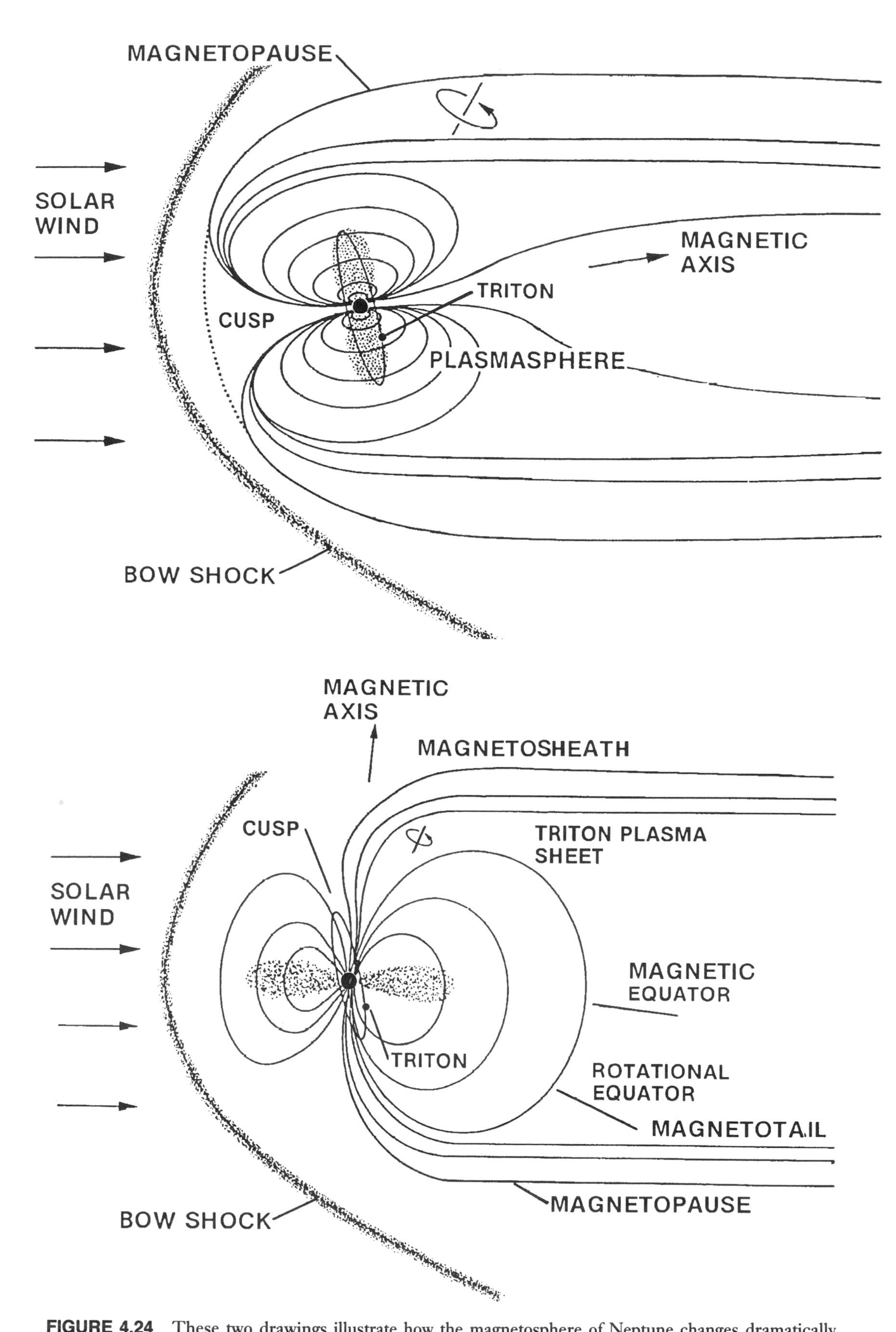

FIGURE 4.24 These two drawings illustrate how the magnetosphere of Neptune changes dramatically as the planet rotates. They show how Triton moves in and out of the radiation belts and how the shape of the magnetotail changes enormously. (After JPL)

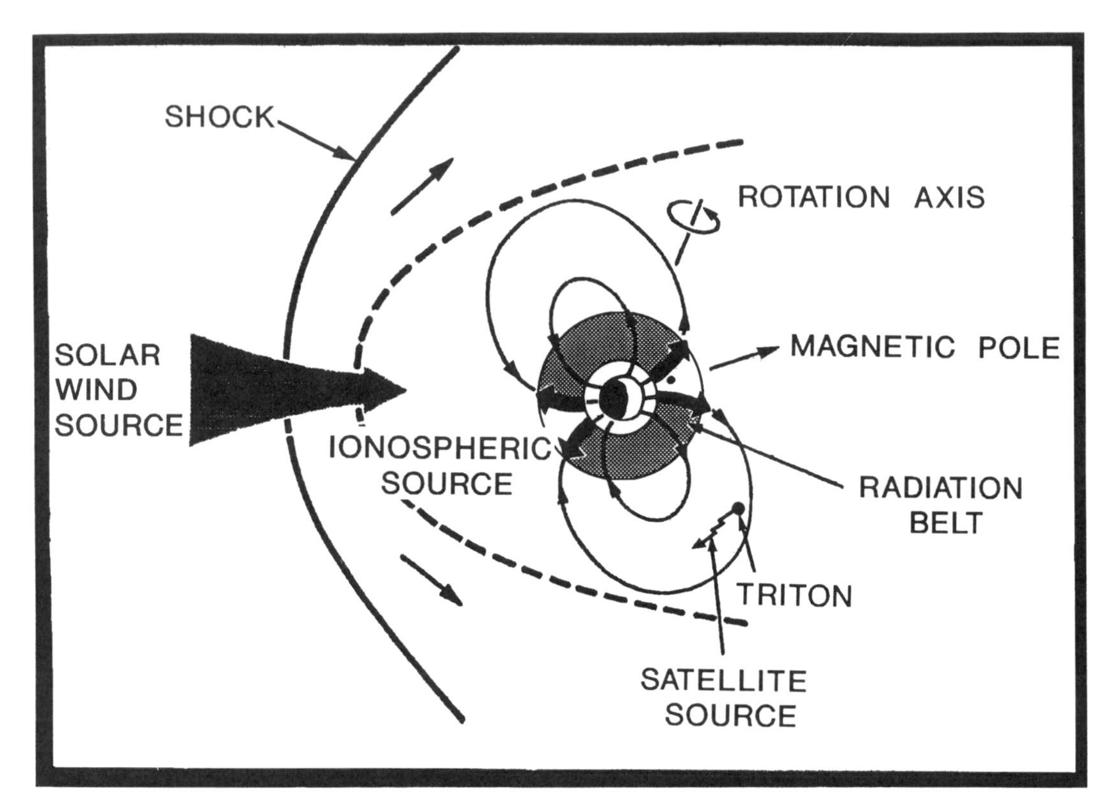

FIGURE 4.25 Sources of energetic particles for the radiation belts are shown in this drawing. These are the solar wind, the planet's ionosphere, and the satellite Triton. (After JPL)

the plasma science experiment (PLS), the low-energy charged-particle instrument (LECP), and the cosmic ray system (CRS). Because the spacecraft entered the magnetosphere toward a magnetic pole of the planet, as mentioned earlier, the transition from the bow shock through the magnetopause and into the magnetosphere was not abrupt but consisted of a gradual change in the density and temperature of the plasma. Based on the location of the bow shock at a distance of 34 planetary radii, a clearly defined crossing of the magnetopause was predicted as likely to occur at about 26 radii from the planet. Instead, the instruments showed a gradual change of the plasma environment with magnetosheath plasma finally giving place to the magnetospheric plasma at a distance of only 23 radii.

Measurements of the flux of plasma by PLS showed increases of both ions and electrons at about 11 planetary radii on the inward leg of the trajectory and at 8.5 planetary radii on the outward leg. These were interpreted as being caused when the spacecraft crossed a sheet or a torus of plasma surrounding the planet. Higher energy electrons were concentrated slightly farther out from the planet than those of lower energy in this sheet or torus. Ions were concentrated closer to the planet than the electrons. The ions, also, appeared to have two major components, those with low energies around 20 volts per charge and those with higher energies around 200 volts per charge.

Two hot plasma regions were also encountered around the time of closest approach of the spacecraft to Neptune. Both were probably plasma sheets. In the outermost region, ion fluxes were as high as 6 kilovolts, and the density of the plasma, assuming it consists of nitrogen ions and protons, was 1.4 particles per cubic centimeter. In the innermost region, ion density and temperature were somewhat less, and the electron temperature much lower than in the outer region. In general, the Neptune plasma is at a very high temperature but at a very low particle density.

Closer to the planet the intensities of trapped electrons and protons decreases as a result of absorptions by the rings and the satellites. The inner satellites produce complexities in the structure of the radiation belts of the planet, and the signature of at least one satellite, 1989N1, was seen in the LECP data. The plasma particle absorption signatures of the rings, Triton, 1989N1, and 1989N2, 1989N3, and 1989N4 were seen in the data from the cosmic ray system. Very few of the plasma particles can originate from the solar wind; most come from the Neptunian system; Neptune, the rings, and the satellites (figure 4.25).

A fast particle impinging on the surface of a satellite ejects atoms and molecules, sometimes at escape velocity. Large satellites retain the particles. There are accordingly processes of erosion and deposition which over millions of years can change several meters depth of surface material. A satellite with an atmosphere, such as Triton, will

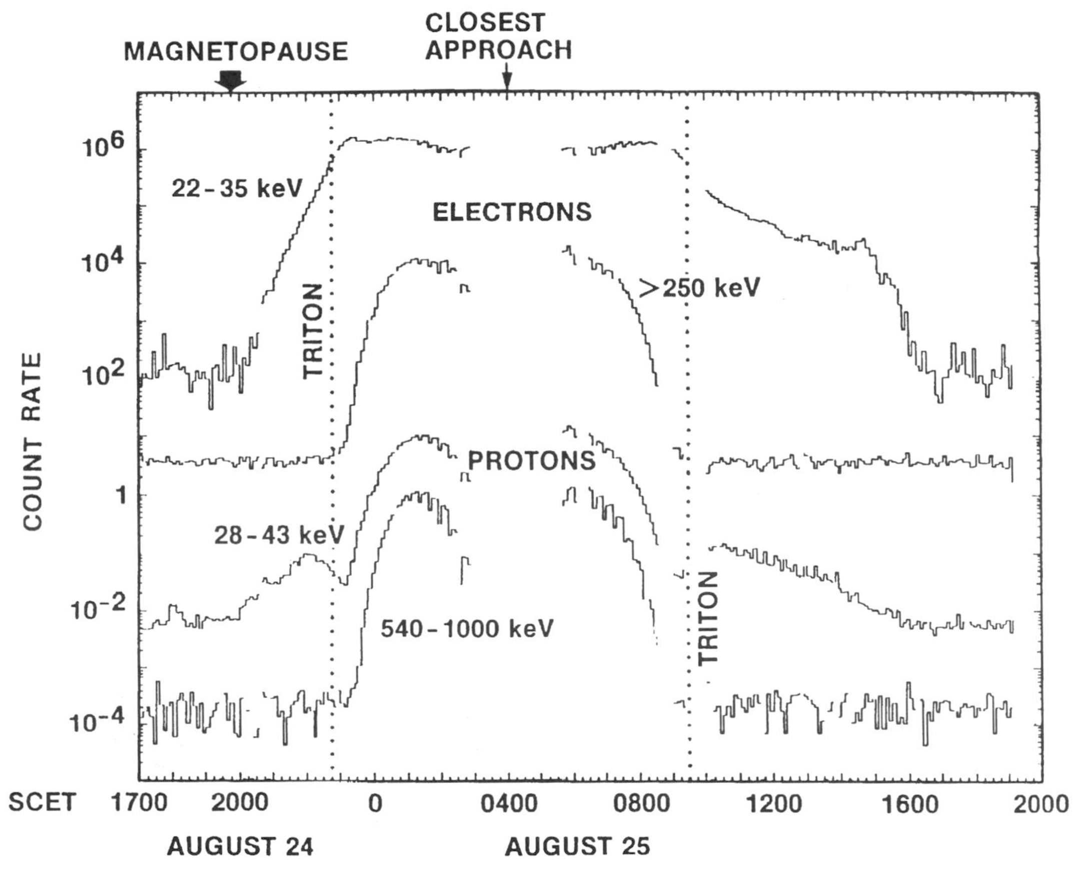

FIGURE 4.26 The Low Energy Particle (LEP) experiment counted energetic particles—electrons and protons—at several energy levels. The radiation belts, evidenced by the upward surge in counting rates on either side of the planet, are within the orbit of Triton which is identified by the dotted vertical lines on the figure. (JPL)

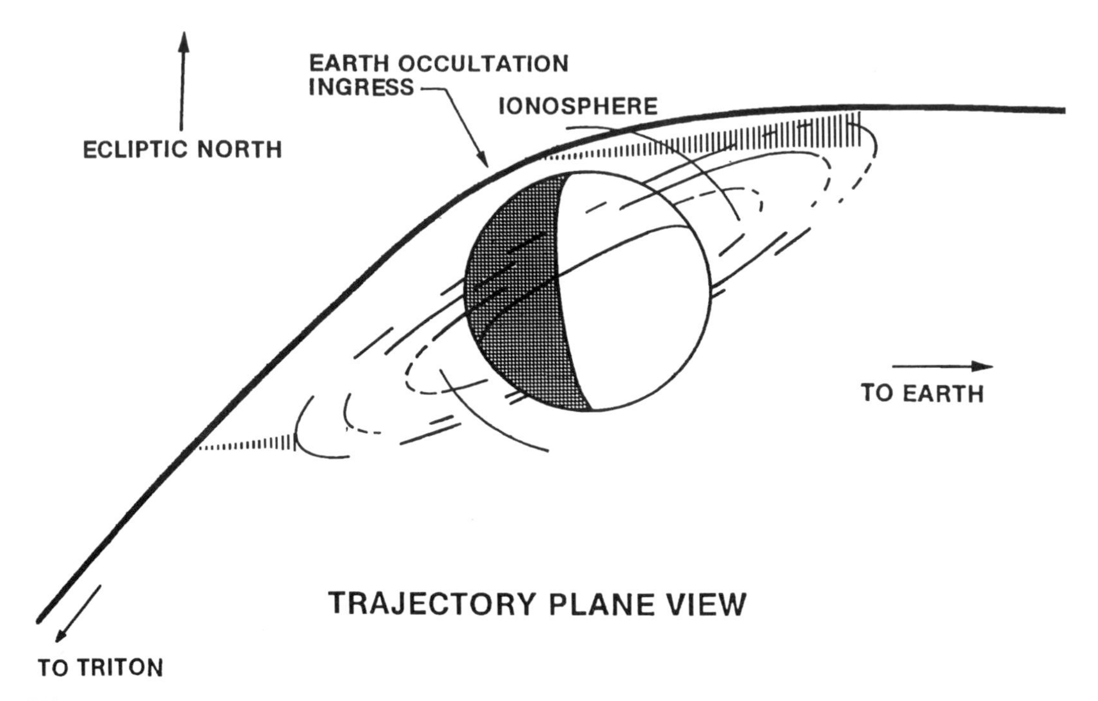

FIGURE 4.27 This drawing of the trajectory of the spacecraft over the north polar region of Neptune shows how radio waves would pass through the rings and then the ionosphere and atmosphere at ingress and again at egress. This was an important radio experiment.

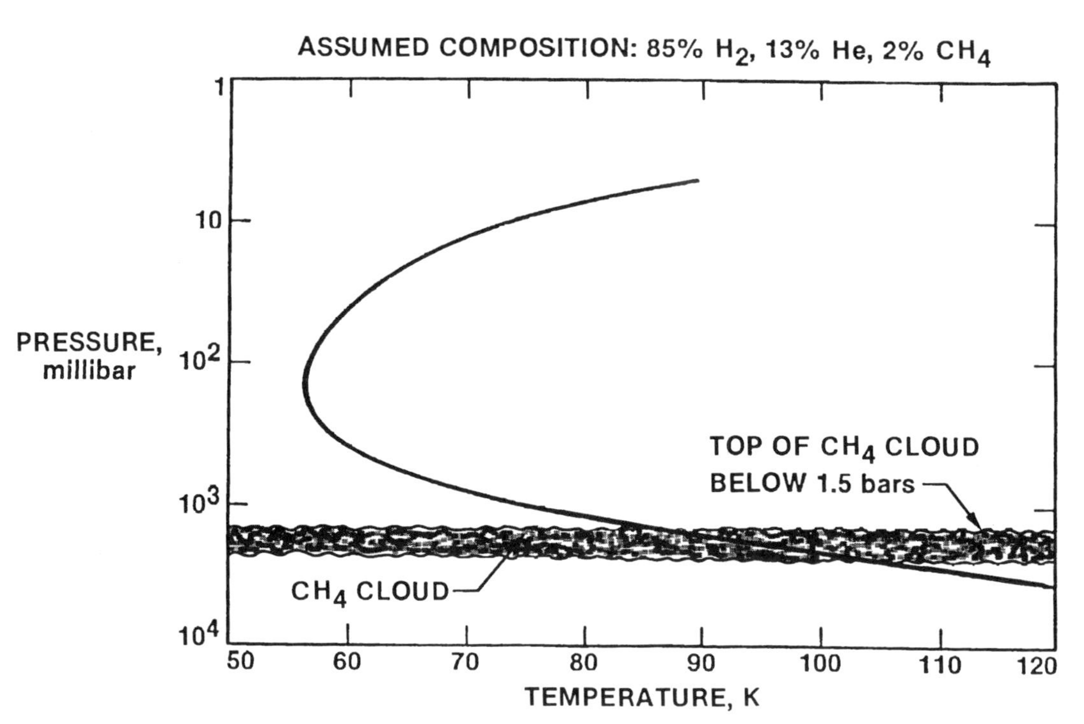

FIGURE 4.28 The temperature of the atmosphere of Neptune was determined from the radio occultation data as plotted on this graph, making assumptions about the composition of the atmosphere. (JPL)

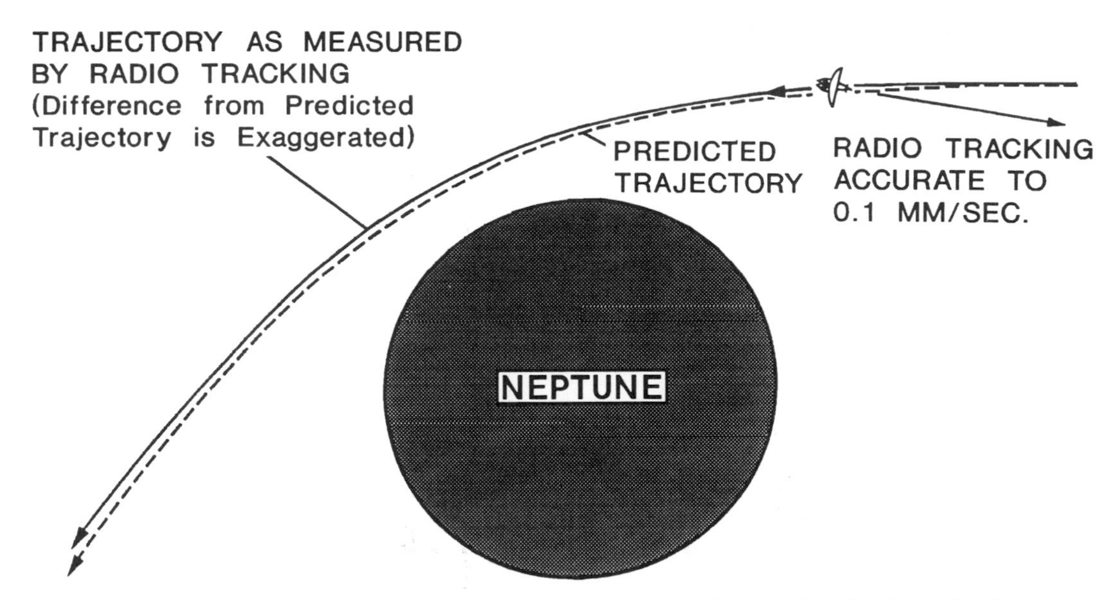

FIGURE 4.29 Another important radio experiment was to use the Doppler shift in radio frequency resulting from the change in velocity between the spacecraft and Earth to plot the trajectory of the spacecraft as it passed Neptune. The difference between the predicted trajectory and the actual trajectory provided information about the gravitational field and the mass of the planet. (JPL)

absorb some particles. Others can, however, become ionized and picked up by magnetic field lines and drawn away from the satellite. This produces a loss of atmospheric gases into the plasmasphere.

However, none of the sources of plasma in the Neptunian system is prolific; Neptune is too cold, the rings and inner satellites have dark surfaces of complex compounds from which ions are not easily sputtered, and Triton has a low atmospheric temperature and its position in the magnetosphere varies enormously as Neptune rotates. These conditions, so very different from the other giant planets, probably play a large part in accounting for the low density of the plasma surrounding Neptune compared with those of the other planets.

The maximum plasma density within the magnetosphere—1.4 plasma particles per cubic centimeter—was smaller than that observed by *Voyager* at any of the other large planets. The counts of energetic particles in the radiation belts of Neptune are shown in figure 4.26 as obtained by the low-energy particle experiment. The radiation belts are within the orbit of Triton, the position of which is indicated on the figure symmetrically about the position of closest approach. Observations of aurora on Triton, as discussed in a later chapter, imply that ions and electrons from the magnetosphere enter the atmosphere of the big satellite. Triton scoops up any particles that stray close to its orbit and as these hurtle into the satellite's atmosphere they produce the auroras detected by *Voyager*'s ultraviolet instrument. The flux of 50 kev particles in the radia tion belts of Neptune is small compared with the other planets, as shown in table 4.4.

There were two types of particles observed within the plasma; light ions of masses between 1 and 5 amu, and heavy ions of masses between 10 and 40 amu. The former probably originate from the atmosphere of Neptune, while the latter most likely come from Triton's atmosphere, ionosphere, or both. Most of these particles are concentrated

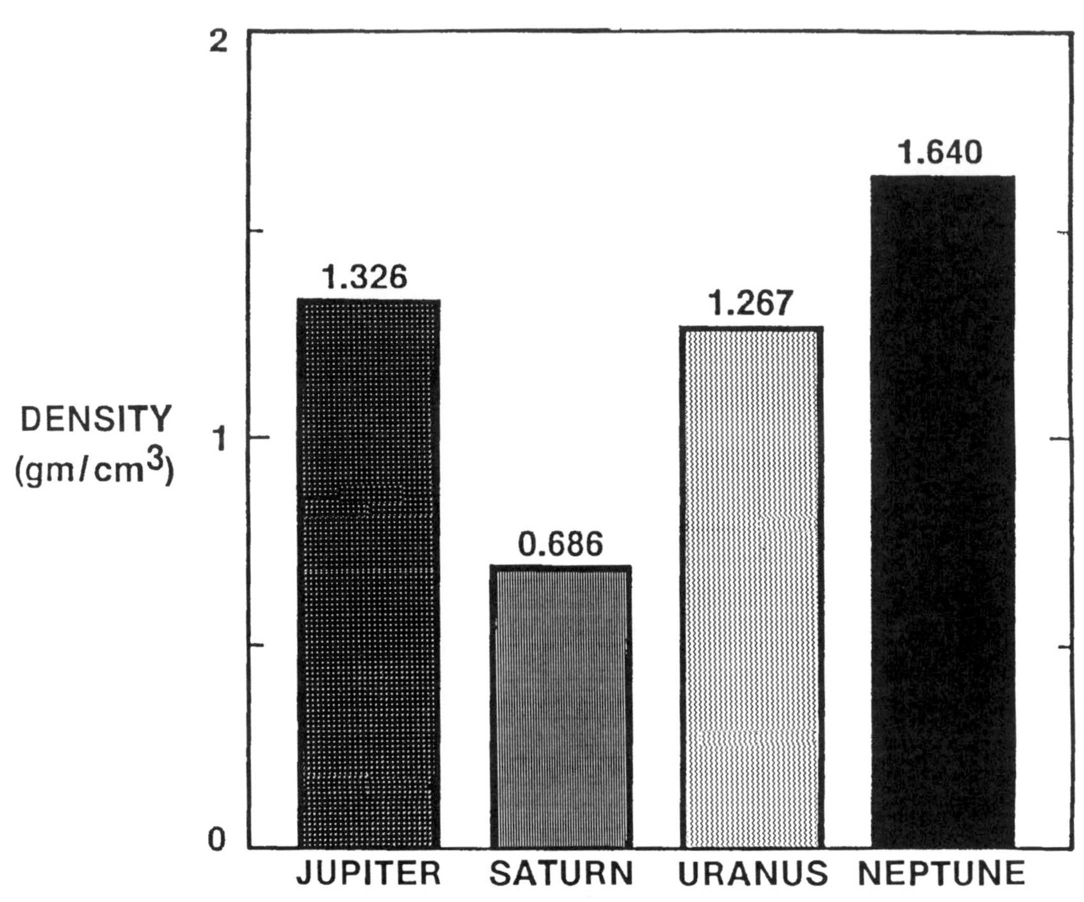

FIGURE 4.30 From the mass of Neptune determined from the radio data, and its diameter determined from the optical data and the radio occultation data, its density could be calculated. Neptune is the densest of the outer planets as shown in this comparison chart.

TABLE 4.4 Radiation Belt Fluxes of 50 kev Particles (Relative to Earth)

Planet	*Ions*	*Electrons*
Earth	100	100
Jupiter	1,000	10,000
Saturn	10	100
Uranus	1	100
Neptune	0.5	10

in a sheet close to the planet. There are also indications of a torus of charged particles. These trapped particles consist of helium, atomic hydrogen, and singly ionized molecules of hydrogen. There are over one thousand times more atomic hydrogen plasma particles than molecular hydrogen ions, and one-tenth as many helium particles as there are molecular hydrogen ions.

A dense plasma torus of nitrogen and methane ions was expected to be associated with Triton analogous to those associated with Io at Jupiter and Titan at Saturn. While some heavy ions were found associated with Triton, there were not as many as expected. This is probably because Triton moves in and out of the plasmasphere so that at times ions are captured within the radiation belts and at other times they escape into the cusps or the magnetotail and are lost to the plasmasphere. This is quite a different situation from that associated with Io at Jupiter where plasma builds up to very high densities. In fact, the magnetosphere of Neptune has the least density of particles of any of the magnetospheres explored by *Voyager*. Even Uranus has almost ten times as many plasma particles as Neptune, as shown in table 4.4.

As *Voyager* flew by Neptune its path was changed by the gravity of the planet. Radio tracking of the spacecraft provided an accurate velocity measurement to 0.1 millimeters per second. For the radio tracking, the spacecraft's transmitters issued two nearly pure signals without telemetry. As the spacecraft traveled behind the planet as seen from Earth (figure 4.27, p. 88), these radio signals passed sequentially through the equatorial plane, the ionosphere, and the atmosphere. As the spacecraft emerged from behind the planet the signals passed once more through the ring plane. Observations of changes to the beam on its arrival at Earth provided information about the features of the rings, the ionosphere and the atmosphere, and also information about the velocity of the spacecraft. For this experiment to be successful the spacecraft had to be timed with great precision. A 16-second error would have ruined the experiment. Actually the spacecraft arrived within 2 seconds of the desired time; a tribute to the skills of the navigation team.

A temperature profile was obtained for the atmosphere between pressure levels of four microbars to 1.5 bars at the latitude of occultation entrance at 60 degrees north. Another profile was obtained for 40 degrees south latitude. The results were similar to those obtained with the IRIS instrument. The temperature profile is shown in figure 4.28, p. 88. The radio occultation also provided updated figures for the dimensions of Neptune. At the one bar level its diameter is 30,776 miles (49,528 km) at the equator, and 30,254 miles (48,680 km) at the poles.

The actual trajectory of the spacecraft as derived from the radio data was compared with the predicted trajectory to obtain a measure of the planet's gravitational field (figure 4.29, p. 89). Coupled with the accurate measurements of the planet's diameter of 30,776 miles (49,528 km) obtained from imaging and radio occultation data, the gravity measurements gave a value for the gravitational constant GM of 6,835,100 km^3/sec^2. From this measurement the density of Neptune is 1.64 grams per cubic centimeter. The densities of the giant planets are compared on the bar chart of figure 4.30. Neptune is the densest member of the group, while Saturn has an inordinately low density of less than that of water.

5

RINGS AND SMALL SATELLITES

Two intriguing questions remained unanswered about the Nepunian system as the spacecraft sped toward the distant planet. Both seemed unlikely to be answered by observations from Earth.

Perhaps the most important was whether the ring system of Neptune consisted of complete rings or of ring arcs or even of rings at all since some theories presupposed that only planets with regular systems of satellites could have ring systems. And all ground-based observations indicated that Neptune most certainly did not have a regular system of satellites. If there were ring arcs only, some theories about how rings form and are maintained would have to be modified, requiring updating of the more intricate principles of celestial mechanics. For example, theory showed that a clump of ring material occupying about 16 miles (20 km) radially from Neptune would spread into a complete ring around the planet in about 5 years only. Such time scales for spreading of material into complete rings made it difficult to imagine how ring arcs could endure without some small satellites holding them in place. The ring system of the planet as derived from Earth-based observations is shown in figure 5.1.

The other question followed naturally whether or not Neptune possessed more satellites than the two observed from Earth. Would the planet have several very small satellites similar to those discovered at the other giant planets explored by the *Voyagers* An isolated observation from Earth had suggested that one other small satellite might

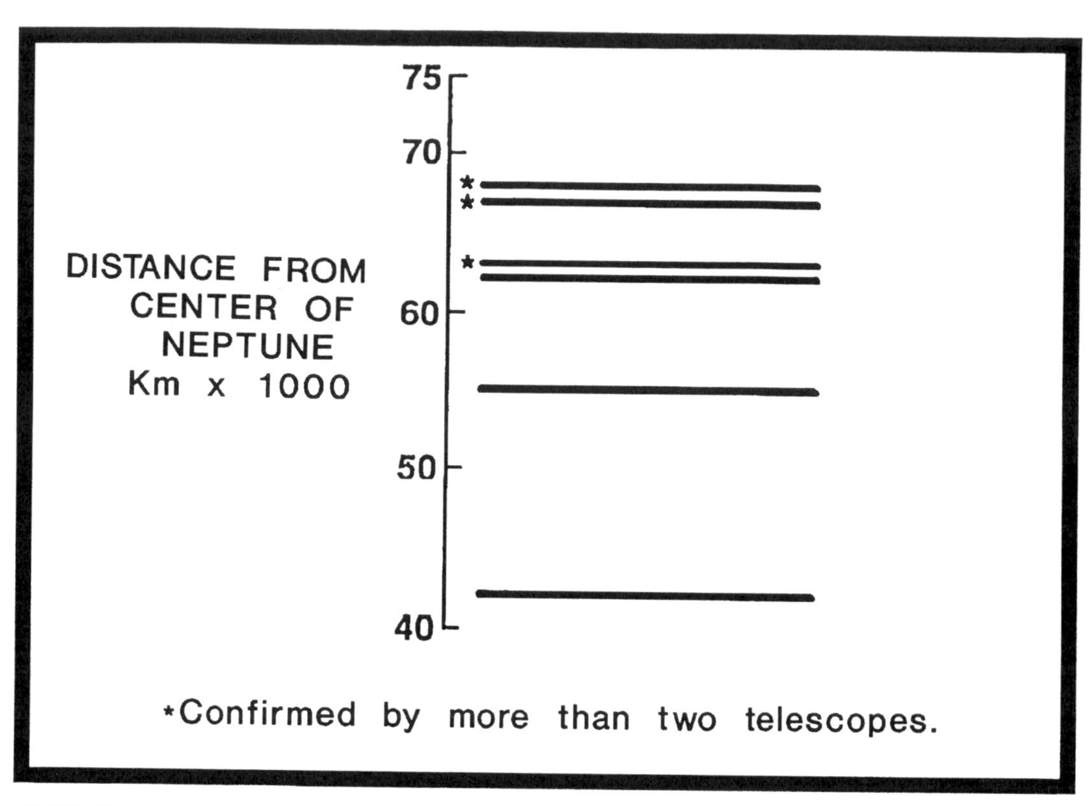

FIGURE 5.1 Distances of rings of Neptune as deduced from observations of stellar occultations by Earth-based telescopes. Those marked with an asterisk were seen by more than two telescopes.

be present, but that observation had been very inconclusive. It depended upon a single obscuration of a faint star in May 1981 that could very well have resulted from a ring arc and not from a small satellite located in an orbit some 47,800 miles (77,000 km) from the planet.

Voyager's encounter with the planet settled the matter. Neptune has rings, ring clumps, and small satellites; but the encounter still left several more detailed and penetrating unanswered questions.

There was a gnawing suspicion as *Voyager* approached its close encounter that there could be danger from particles within and around these elusive rings. Estimates had been made, based on the terrestrial observations, that there were at least two nearly circular partial ring arcs located about three planetary radii out from the planet (see figure 5.1) close to the equatorial plane. The arcs were about 5 to 12 miles (8 to 20 km) wide and contained particles ranging in size from dust to pebbles. The chosen flight path of *Voyager* would carry the spacecraft through the ring plane close to the ring arcs, and mission planners were worried about the uncertainties in locating the positions of the ring arcs from the terrestrial observations. Passage through diffused sheets of particles could result in damage to the spacecraft, especially to its optical systems as it passed through the ring arc plane, which was assumed to be in or close to the equatorial plane. As mentioned in chapter 3, they decided to orient the spacecraft so that the optical surfaces would be protected from small particles, but this would not prevent general damage to the spacecraft from impacting objects the size of pebbles. The path of *Voyager* lay beyond the outermost ring arc as derived from the stellar occultations observed at Earth, and the path could be moved further out if images obtained by *Voyager* should prove this to be necessary. But the final choice of flyby had to be no later than one week before closest approach. As *Voyager* speeded toward the planet, excitement grew rapidly.

Voyager's imaging system did not confirm the presence of the ring arcs until August 11, 1989 when an extremely faint image of an arc was apparent on an image frame showing several newly discovered small satellites of Neptune. The spacecraft was at a distance of 13 million miles (21 million km) from the planet. The arc was just outside the orbit of satellite 1989N4 which moves around Neptune at a radial distance of 38,500 miles (62,000 km). The arc covered 45 degrees of a circular orbit, about 30,000 miles (50,000 km) in length, and was later determined to be at a radial distance of 39,100 miles (62,930 km). The arc (figure 5.2) was named 1989N1A. Another arc was discovered trailing behind satellite 1989N3 by approximately 90 degrees. This satellite is in an orbit with a radius of 32,300 miles (52,000 km). Later it was determined that the ring arc, tentatively labeled 1989N2A, was at a distance of 33,000 miles (53,200 km). So far, neither the ring arcs nor the small satellites corresponded closely with ground-based observations, except 1989N1A which was very close to the radius of ring arcs determined from occultation observations made in June and August 1985. Following these discoveries by *Voyager,* scientists speculated that the ring arcs might consist of debris associated with the nearby small satellites or even the remnants of satellites that had been broken apart by collisions.

After the excitement of confirming the presence of the ring arcs, scientists became mystified when subsequent images did not show the shorter arc. It had vanished from the images. However, on August 22, Bradford Smith, leader of the imaging team, was able to announce to the press that not only had the lost arc been found again, but also subsequent images showed that it was part of a complete ring.

The plasma wave subsystem (PWS) was not originally intended to detect dust particles but it turned out that the impact of particles on the body of the spacecraft caused miniature explosions as kinetic energy was converted into heat and produced momentarily very high temperatures and ionized gases. In turn, these conditions caused voltage pulses that the PWS could record. D. A. Gurnett, principal investigator for the experiment, had seen this impacting effect at Saturn. Now he looked for it at Neptune to serve as a measure of danger to the spacecraft as it crossed the ring plane.

"We started detecting dust particles about two hours before the ring plane crossing." He told the press. "Watching our data display in real time, we found it very scary as the impact rate went up, and continued for about 10 minutes."

The impact rate rose dramatically to a maximum of about 300 particles per second at about 53,000 miles (85,500 km) from the center of the planet, before the danger passed and the ring plane crossing had been made safely. These smoke-sized particles appeared to be concentrated densely in a disk. But "dense" in this environment meant that there were about three particles only for each 1000 cubic meters of space. Later, dust particle impacts were also recorded by the PWS instrument for two hours after the outbound crossing of the ring plane which occurred at 64,400 miles (103,700 km) and was at a shallower angle than the inbound crossing (figure 5.3). The peak impact rate at the outbound crossing was about one-third that at the inbound crossing. Also, particles were detected over the polar region during the closest approach to Neptune,

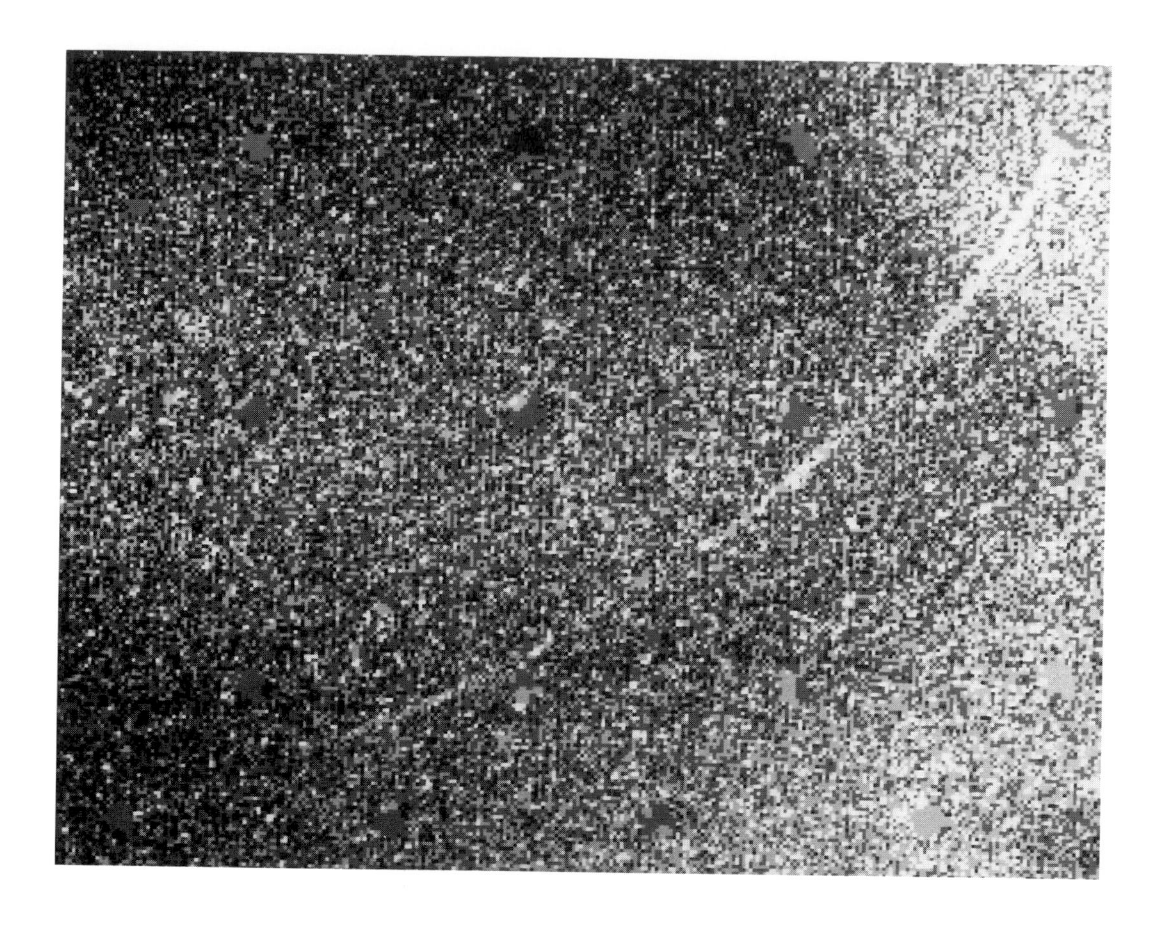

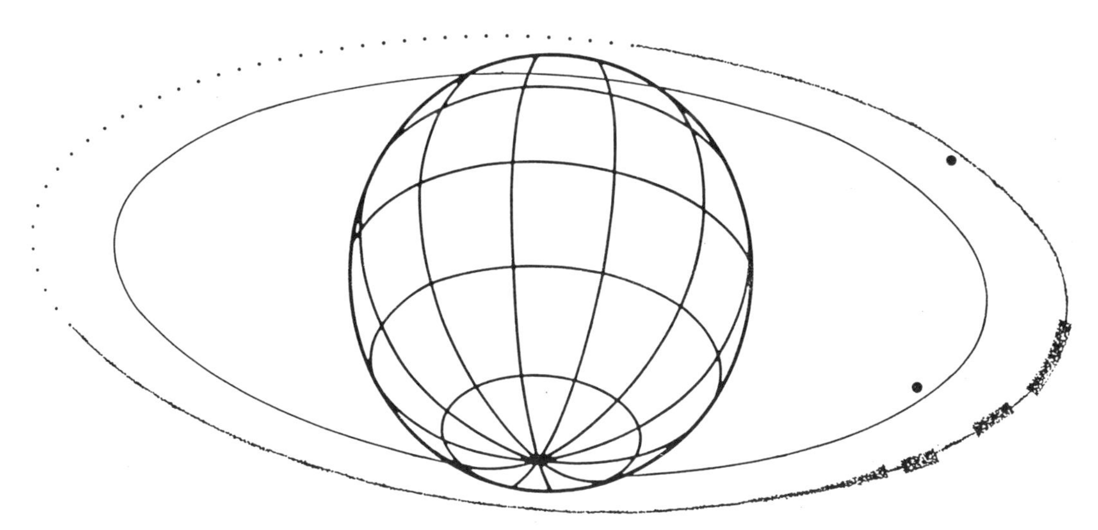

FIGURE 5.2 The *Voyager* spacecraft was 5.3 million miles (8.6 million km) when it took this 61-second exposure picture in search of faint rings and small satellites. The portion of a ring arc recorded in this picture is 35 degrees in longitudinal extent. It has three segments separated from each other by about five degrees. The ring arcs and two newly discovered small satellites are shown on the drawing beneath the photoimage. (NASA/JPL)

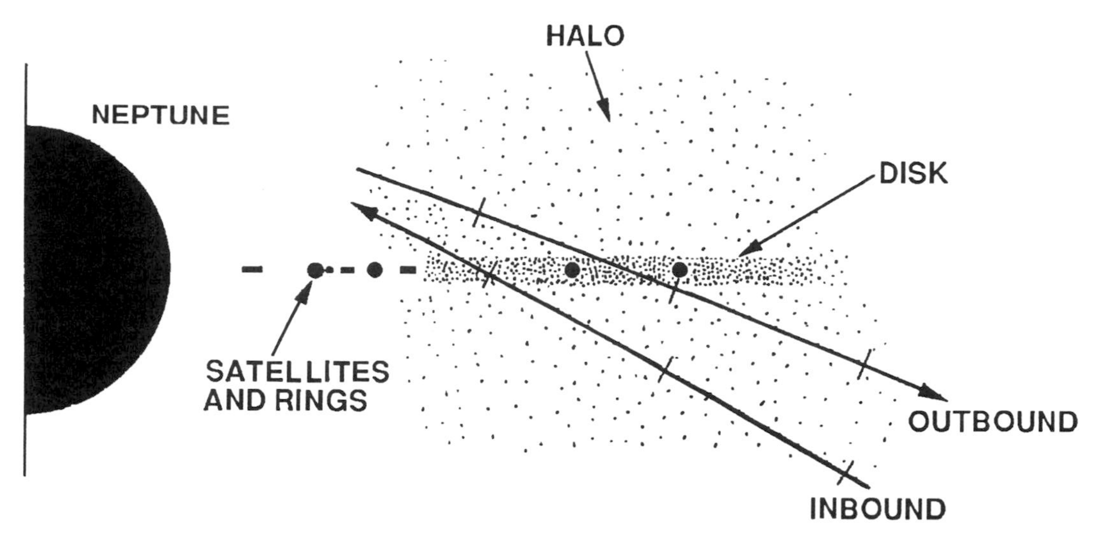

FIGURE 5.3 The Plasma Wave Subsystem (PWS) although not originally intended to do so was able to detect the impact of dust particles on the structure of the spacecraft which produced voltage impulses. The data from this instrument recorded a disk of particles in the plane of the rings and a halo of particles above and below that plane. Data were obtained on both the inbound and outbound legs of *Voyager*'s trajectory. (JPL)

and at 12,000 miles (20,000 km) above and below the plane of the rings. As mentioned earlier there was speculation that Neptune might have rings that pass over the poles of the planet produced by Triton's perturbations on ring particles. Such polar rings would be perpendicular to the plane of Triton's orbit. The particles observed over the polar regions could be evidence of a very diffuse polar ring, but such a ring was not apparent in any other data returned from the spacecraft. The thick cloud of dust outside the plane of the rings is most likely caused by the tilted magnetic field spreading ionized particles from within the ring system.

On closer approach to the planet the spacecraft returned images at higher resolution. Computer enhancement of these images showed that the arcs are, in fact, parts of fainter rings that go all around the planet. The rings proved to be so diffuse, and the material particles in them so small and dark, that the full rings could not be detected from Earth. It turned out that the three thicker segments of the brightest ring accounted for all the observations of ring arcs made from Earth; except one. This unique exception was, so it seems, the incredibly coincidental occultation by a small satellite, the newly discovered 1989N2, whose orbit's radius is 45,700 miles (73,600 km), which might have been 1981N1 whose orbital radius had been estimated as 47,800 miles (77,000 km). As detailed in the author's previous book *Uranus and Neptune,* the 1981 occultation data indicated the presence of a satellite some 125 to 180 miles (200 to 300 km) in diameter, and 1989N2 is about 125 miles (200 km) in diameter. The mystery of the disputed third satellite of Neptune had now been solved. Three satellites had, in fact, been discovered before *Voyager* reached the planet.

Two images taken on August 26 from a distance of 175,000 miles (280,000 km) showed two main rings and an inner, fainter ring at about 25,000 miles (42,000 km) from the center of the planet. A faint band extended smoothly from the 33,000-miles (53,000-km) radius ring to roughly half way between the two bright rings. These newly

FIGURE 5.4 This wide angle picture of the shadow of Neptune on the ring system was taken when the spacecraft was 412,000 miles (660,000 km) from the planet. The shadow clearly delineates the radial extent of the broad sheet of material extending from about half-way between the two bright rings down to, and perhaps beyond, the inner bright ring. (NASA/JPL)

discovered rings are much broader and fainter than the bright rings imaged earlier. These images were taken from the far side of Neptune after the closest encounter. They showed up because they were backlit, with the Sun shining from behind them. Large particles scatter light backward toward the source whereas small particles scatter the light forward, away from the source.

This backlit lighting condition of the rings enhanced the visibility of dust with the consequence that the extremely faint dusty rings could be recorded on the image frames. Even so, they were so faint that long exposures had to be given to detect them. The two main rings were also shown to be complete rings. In the backlit image the shadow of the planet could be clearly seen stretching across the whole ring system indicating the presence of the faint but extensive dust sheet. There was now no doubt that the rings are complete and not just arcs. A comparison of the brightness of a ring as imaged in forward scattered and backward scattered light provides information about the sizes of the particles in the ring.

The material within the rings varies considerably in size from ring to ring. The largest proportion of the fine material, about the size of particles in smoke, occupies a region of a broad diffuse sheet outside the innermost ring as revealed by the shadow of the planet on the ring system (figure 5.4). The rings of Neptune have much less large material than those of Uranus and they were not readily seen in the radio occultation data. But they have much more dust. There may be rocks as large as 6 to 12 miles (10 to 20 km) in diameter embedded in the rings at the dusty arcs where the outer edges appear brighter. Alternatively, these brighter parts may be merely accumulations of smaller particles. The arcs do, however, appear to have more dust than in the rest of the 1989N1R ring which, in turn, has less dust than 1989N2R and 1989N3R. The orbital locations of the rings relative to those postulated from Earth-based occultation data and the newly discovered small satellites are shown in figure 5.5.

A preliminary nomenclature was developed late during the encounter. The main ring, (officially known as 1989N1R following the convention established by the International Astronomical Union) is at a radius of 39,000 miles (62,900 km) from the center of Neptune. The main ring contains three separate regions where the material is brighter and denser, and explains most of the sightings of ring arcs. These arc sections are not, however, located equidistantly around the ring but are concentrated within 33 degrees (figure 5.6). Several images show what appear to be clumps of very small particles, or of discrete larger bodies, embedded within the rings. Good images of the arcs were obtained in both backward scattered and forward scattered light. These images gave clear evidence of azimuthal structure of about 6 miles (10 km) size. If these were satellites they would be crescent phased and less bright in the forward scattered

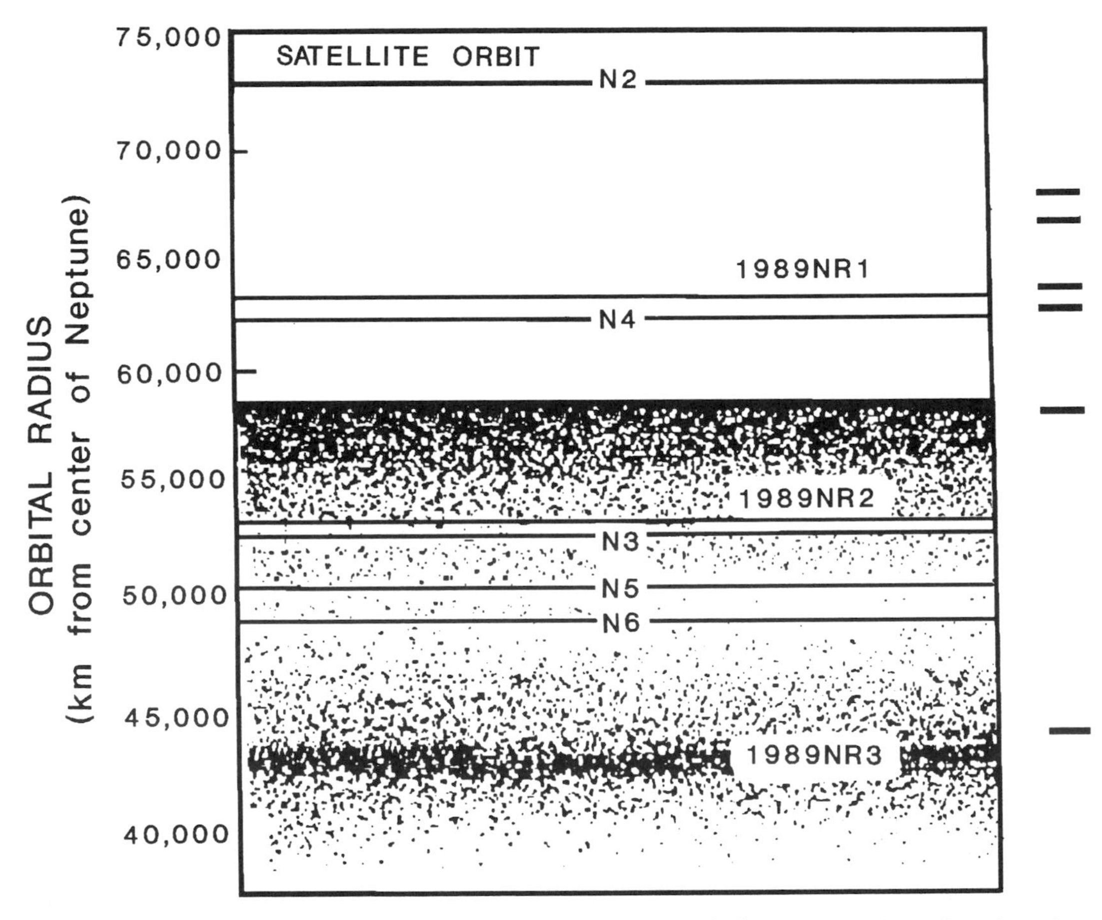

FIGURE 5.5 The radial distances of rings of Neptune and the newly discovered small satellites from the center of Neptune are shown in this diagram. At the right side the dashes show the radial distances of the rings suggested by the star occultation observations. (After JPL)

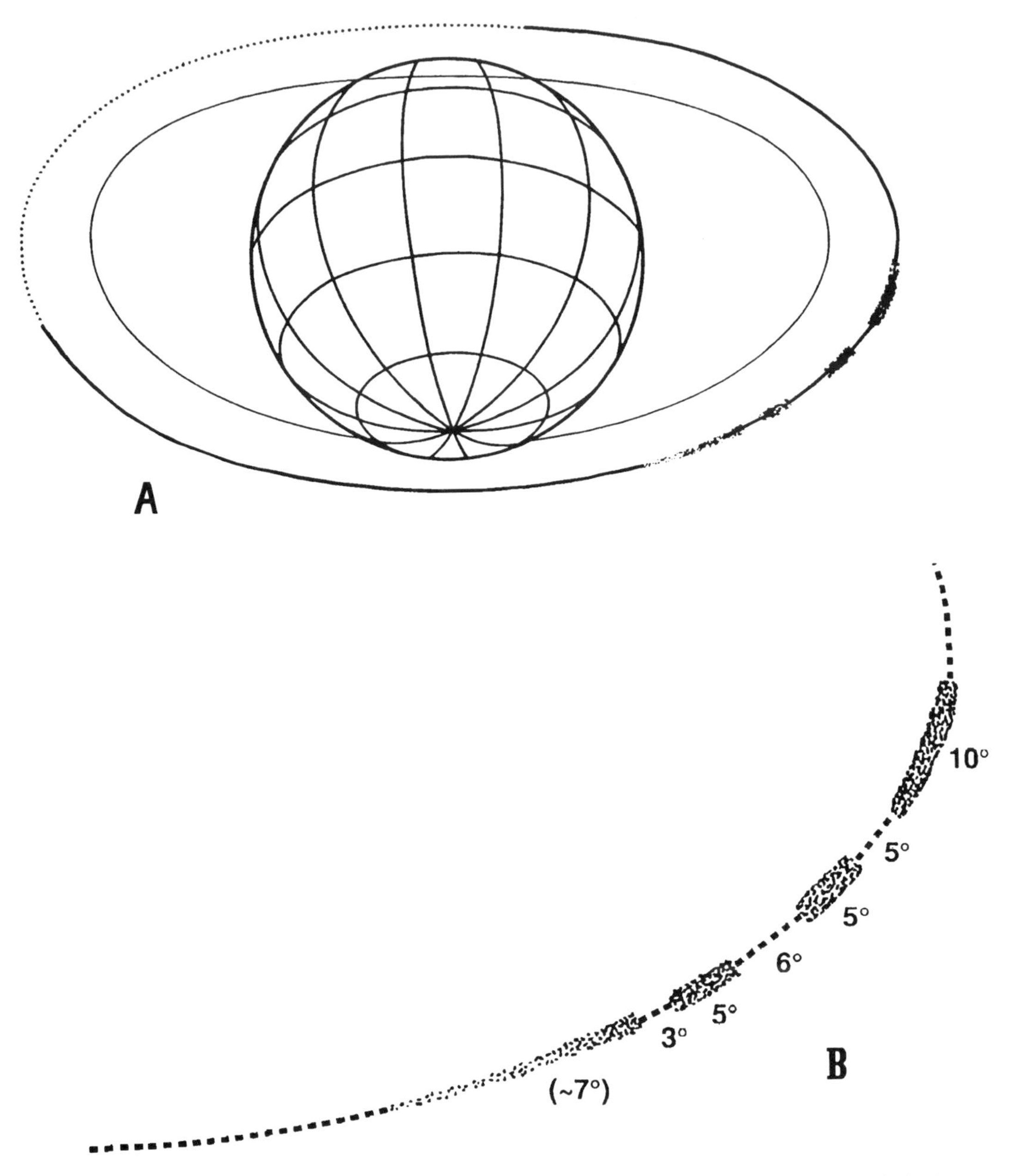

FIGURE 5.6 (A) The ring arcs mapped by *Voyager*'s cameras are shown relative to the ring of which they form a part. (B) The angular size and separation of the ring arcs are shown in this diagram.

condition. That this was not true implies that they must consist of aggregates of particles, perhaps surrounding larger bodies.

How these arcs remain in the rings is still unexplained. Several theories have been suggested but none offer satisfactory explanations. Moreover, no satellites have been found at locations where they might act as shepherding satellites. In fact, despite searches of the images, no satellites larger than about 7.5 miles (12 km) in diameter have been found within the ring system. The small satellite 1989N6 has almost a

FIGURE 5.6 (C) This image with an exposure of 111 seconds was the first to show the rings in detail. This image was obtained after the spacecraft passed the planet. The rings are backlit and show up more clearly. Also the brightness of the rings when backlit means that they contain much dust. (JPL)

corotation resonance with the ring arcs but it is too small and too far away to cause the arcs. The imaging team concluded that the arcs in the 1989N1R are the same as those observed in the occultations seen from Earth and, therefore, that the arcs have existed for at least five years. However, there is evidence that some observations from Earth were of arcs that are no longer present. These arcs seen from Earth were at locations where at the time of the encounter of *Voyager* there were only broad diffuse areas of dust. If the arcs were formed by a satellite being disrupted by impact with a large meteoroid they would be expected to last for only a decade before the material would be distributed all around the orbit. It appears most unlikely that *Voyager* encountered Neptune at precisely the right time to see ring arcs before they dissipated. Continued observations of these arcs is needed to decide if this theory of formation holds.

An inside diffuse ring (1989N3R) is a complete ring located about 26,000 miles (41,900 km) from the center of the planet. This ring might extend as a very diffuse sheet all the way down to the atmosphere where its particles become meteors. However, this cannot be established with certainty from the imaging data because the amount of scattered light is unknown. The innermost ring of Saturn (D-ring) extends down to the atmosphere in forward-scattered lighting conditions, as does the innermost ring of Uranus (1986U2R).

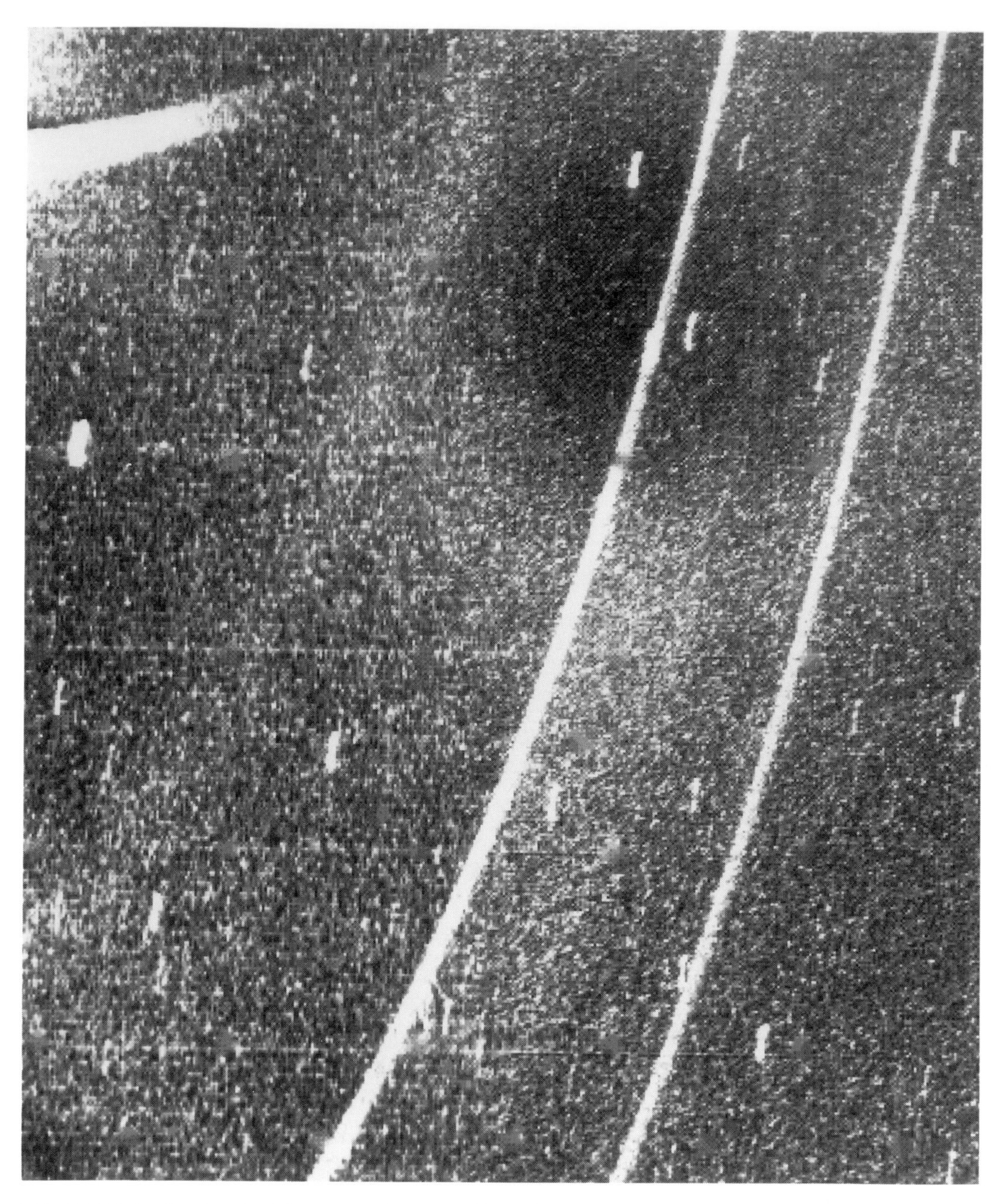

FIGURE 5.7 Another backlit image shows details in the two bright rings and a faint sheet of material extending from a faint ring. Another faint broad ring was discovered in this image. (NASA/JPL)

Neptune's innermost of the major rings, designated 1989N2R, has a radius of about 33,000 miles (53,200 km).

Another broad ring (1989N4R) covers a radial distance of 33,000 to 36,600 miles (53,200 to 59,000 km) (figure 5.7). This area was initially called "the plateau." It is a broad diffuse sheet of extremely fine material just outside the innermost major ring, 1989N2R. Generally the broad ring does not contain much dusty material; only about

10 percent. However, in forward scattered light the outer edge of this plateau has a bright edge which suggests that there is a concentration of dust there.

A faint discontinuous ring (1989N5R) was discovered in the backlit images. This ring is on the outer edge of the plateau at a radial distance of 35,730 miles (57,500 km) and may, indeed, be a belt of small satellites.

An extremely faint and narrow ring (1989N6R) is detected in the backlit images at a distance of 38,530 miles (62,000 km), which is the radius of the orbit of the small satellite 1989N4.

Because *Voyager* was able to observe the rings over a wide range of angles of illumination (phase angles), scientists were able to calculate the size of material in the rings. All the regions of Neptune's ring system (figure 5.8) are brighter at a high phase angle (backlit) and this leads to the conclusion that the rings are extremely dusty.

While the dusty areas of Neptune's ring system are extremely faint on the images returned from *Voyager,* the amount of dust in them is appreciably greater than that in the faint rings of Jupiter, more substantial even than the dust bands of Uranus. The question is how this population of dust is maintained since dust particles would be expected to be lost to a ring system more rapidly than large particles. Because there

FIGURE 5.8 Two 591-second exposure images have been brought together to produce this general view of the backlit ring system from a position 175,000 miles (280,000 km) beyond Neptune. The bright glare is due to overexposure of the crescent of Neptune. All the rings are visible on the images, including the faint broad rings. (NASA/JPL)

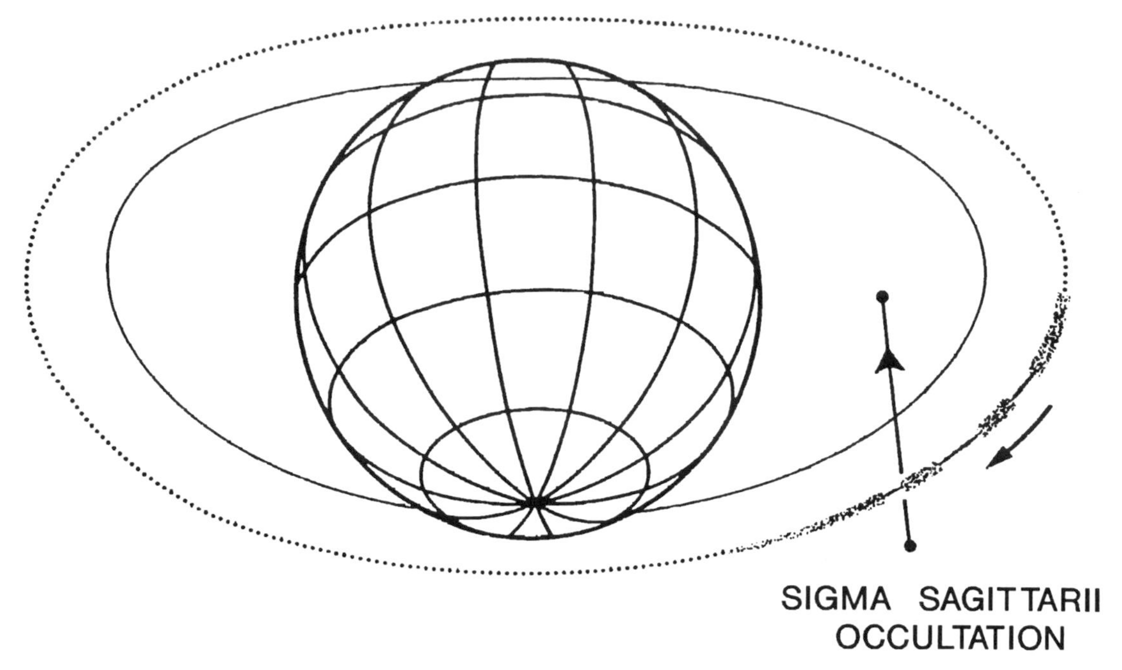

FIGURE 5.9 An important observation was the occultation of the star Sigma Sagittarii by one of the ring arcs. The track of the star through the system is shown in this drawing from JPL.

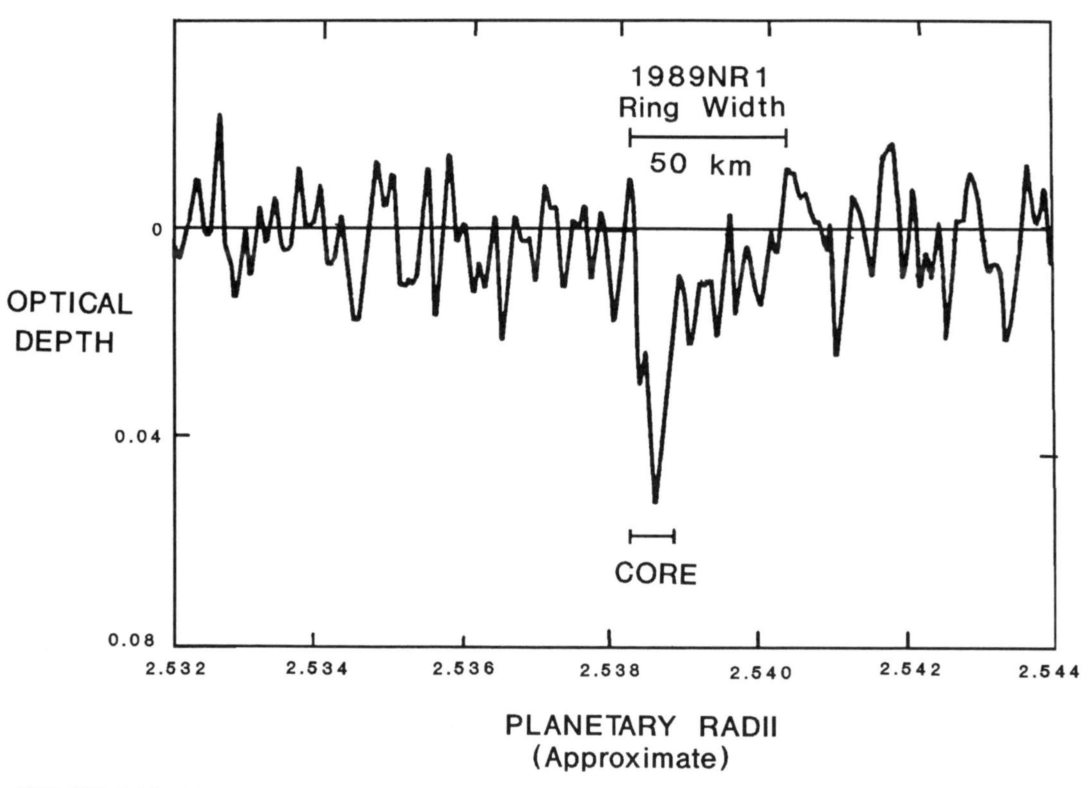

FIGURE 5.10 Variations in the light from the star as it passed behind the rings were recorded by the photopolarimeter and produced the curve shown in this graph. The ring width is shown to be about 30 miles (50 km) with a dense core located close to the inner edge. (After JPL)

may not be a sufficient population of meteorites far out in the Solar System to replenish the dust particles at Neptune in the same way that Jupiter and Saturn gather meteoric dust into their rings, scientists have speculated that there may be large bodies in the Neptunian ring system. These would, however, not be large enough to be imaged as satellites, but their destruction by collisions could provide a source of dust to replenish that lost from the rings to Neptune's atmosphere. If such bodies were present at Neptune in very thin ringlet structures, as were seen at Saturn and Uranus, they could not be detected on the Neptune images because of the long exposure times required so far away from the Sun. The long exposures that were made smeared the images and prevented resolution of fine details that might have revealed any ringlets present in the Neptunian rings.

Important information about the rings was obtained by the occultation experiments. These were of two kinds, an optical occultation of Sigma Sagittarii by the rings, and a radio occultation. The optical occultation (figure 5.9) was recorded by the photopolarimeter to produce the profile shown in figure 5.10. Before the occultation the star had been observed by the photopolarimeter to ensure that its light was not varying and variations during the occultation could be accurately attributed to the presence of ring particles. The starlight passed through the 1989N1R ring just within one of the bright arcs. The ring segment had a core close to the inner edge of the 30-mile (50-km) wide ring, and in this respect the data were similar to Saturn's F-ring and some Uranian rings which also have cores and clumps of particles. Unfortunately the radio occultation geometry was not favorable for obtaining good data, only the outermost ring was suitably placed.

The dimensions of the ring system of Neptune are given in table 5.1. Estimated percentages of dust within the various rings are also given. By comparison the F-ring of Saturn has about 70 percent dust. The ring systems of the large outer planets are compared in figure 5.11, which shows their relationship to the Roche limit and the accretion limit for each planet and to the various small satellites. The Roche limit is the radial distance from a planet within which the survival of a satellite depends on its internal strength (which depends on its composition), and accretion of particles into large bodies greater than a few miles in diameter is inhibited. The accretion limit is the distance from the planet within which it is extremely difficult if not impossible for a satellite to grow by accretion of other small bodies. Most of the rings of the planets generally are located between these two limits, and the satellites of each planet decrease in size the closer each is to the planet. The number of small satellites in the Uranian

TABLE 5.1 Dimensions of the Ring System of Neptune

		Orbital Radius	
Ring Name	*Character*	*miles*	*kilometers*
1989N3R	Diffuse, 40–60% dust	26,000–30,450	41,900–49,000
1989N2R	40–60% dust	33,000	53,200
1989N4R	Faint, 10% dust	33,000–36,600	53,200–59,000
1989N5R	Discontinuous	35,730	57,500
1989N6R	Faint, narrow	38,530	62,000
1989N1R	Ring 30% dust	39,199	62,930
	Ring arcs 60% dust	39,199	62,930

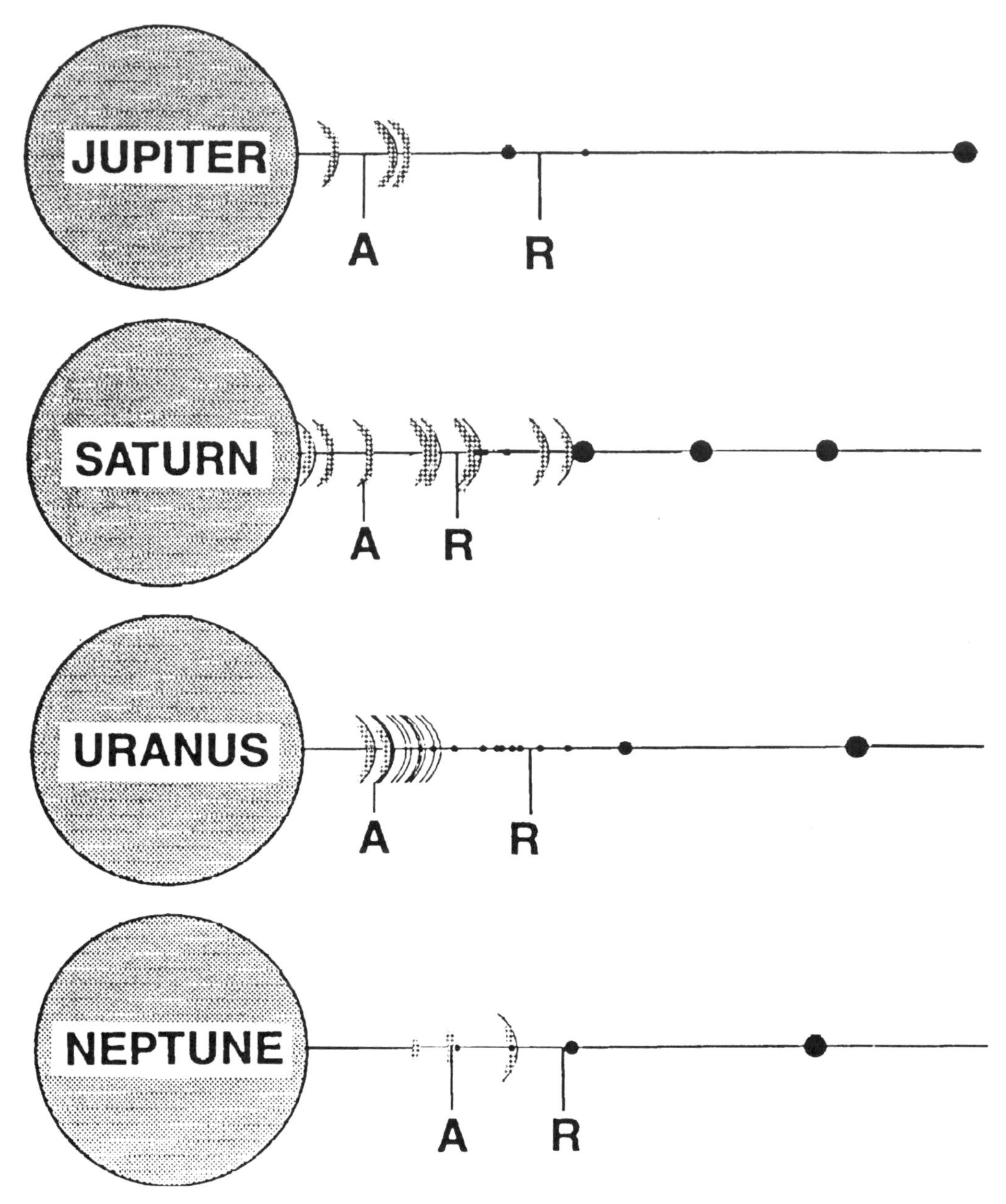

FIGURE 5.11 The ring systems and orbits of small satellites of the giant outer planets are compared in this drawing. Distances are expressed in terms of planetary radii. The letters A and R refer to the accretion limit and the Roche limit described in the text.

and Neptunian systems close to the ring systems are comparable. However, the area of the rings of Uranus is one hundred times greater than the area of the rings of Neptune, and presumably the amount of material in the Neptunian rings is very much less than that in the rings of Uranus. It could be that the greater area of the Uranian rings is evidence of a much more recent breakup of a small satellite in the Uranian system than has occurred at Neptune.

It seems certain that the rings and the small satellites of the big outer planets are dynamic systems in which the satellites break up to form the rings, for although the planetary rings look very different, they have, in fact, many similarities. Saturn's magnificent system contains rings that are similar to those in both the Uranian system and the Neptunian system. Saturn's system seems to have examples of all the different types of planetary ring and forms an important system to study. The rings of Saturn are the brightest, which is probably because the temperature there is too high for the rings to contain much methane. By contrast the rings of Uranus and Neptune are extremely dark, almost as black as soot. They probably contain much methane which has been modified by radiation to produce dark coatings of hydrocarbons. Detailed structure within the Saturn system is more easily imaged because the rings are not only brighter but also receive more sunlight than those of Uranus and Neptune. The Cassini mission to Saturn planned for the late 1990s will accordingly be a most important mission from the standpoint of gaining a better understanding of planetary rings; their formation, dynamics, and dissolution.

In mid-June 1989 inspection of a 46-second exposure image returned by *Voyager* showed a smudged spot of light. Within a few days, subsequent images showed other positions of this object so that an orbit could be calculated. *Voyager* had discovered a new satellite of Neptune (figure 5.12). Its diameter was at that time uncertain; somewhere between 125 and 400 miles (200 to 600 km). Its orbit was close to a circle in the equatorial plane of Neptune and about 73,100 miles (117,650 km) from the center of the planet. The orbital period was 26 hours 56 minutes. The newly discovered satellite was designated 1989N1 until a permanent name can be given by the International Astronomical Union. This small satellite could not be seen from Earth because it is so close to Neptune that it is hidden in the glare of the planet. The tiny satellite was outside the ring arcs seen from Earth and its presence gave mission planners added confidence that the spacecraft could survive any radiation environment of Neptune because the satellite would sweep charged particles from such belts.

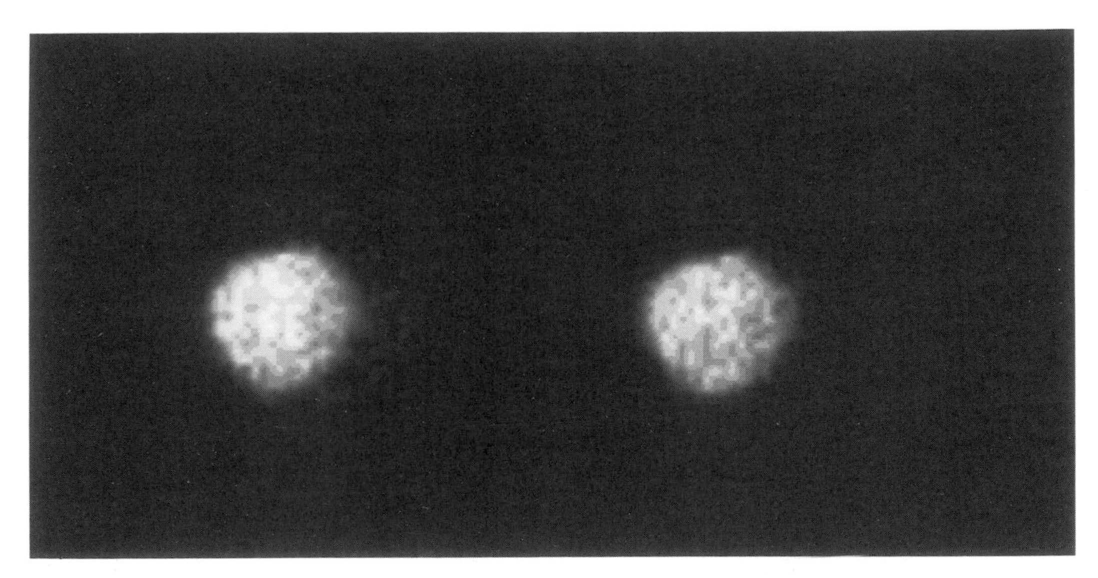

FIGURE 5.12 (A) These two clear filter images show the largest of the newly discovered satellites of Neptune, 1989N1. The satellite's surface is extremely dark, reflecting only six percent of incident sunlight. The satellite is irregularly shaped with a mean diameter of 250 miles (400 km), slightly larger than Nereid.

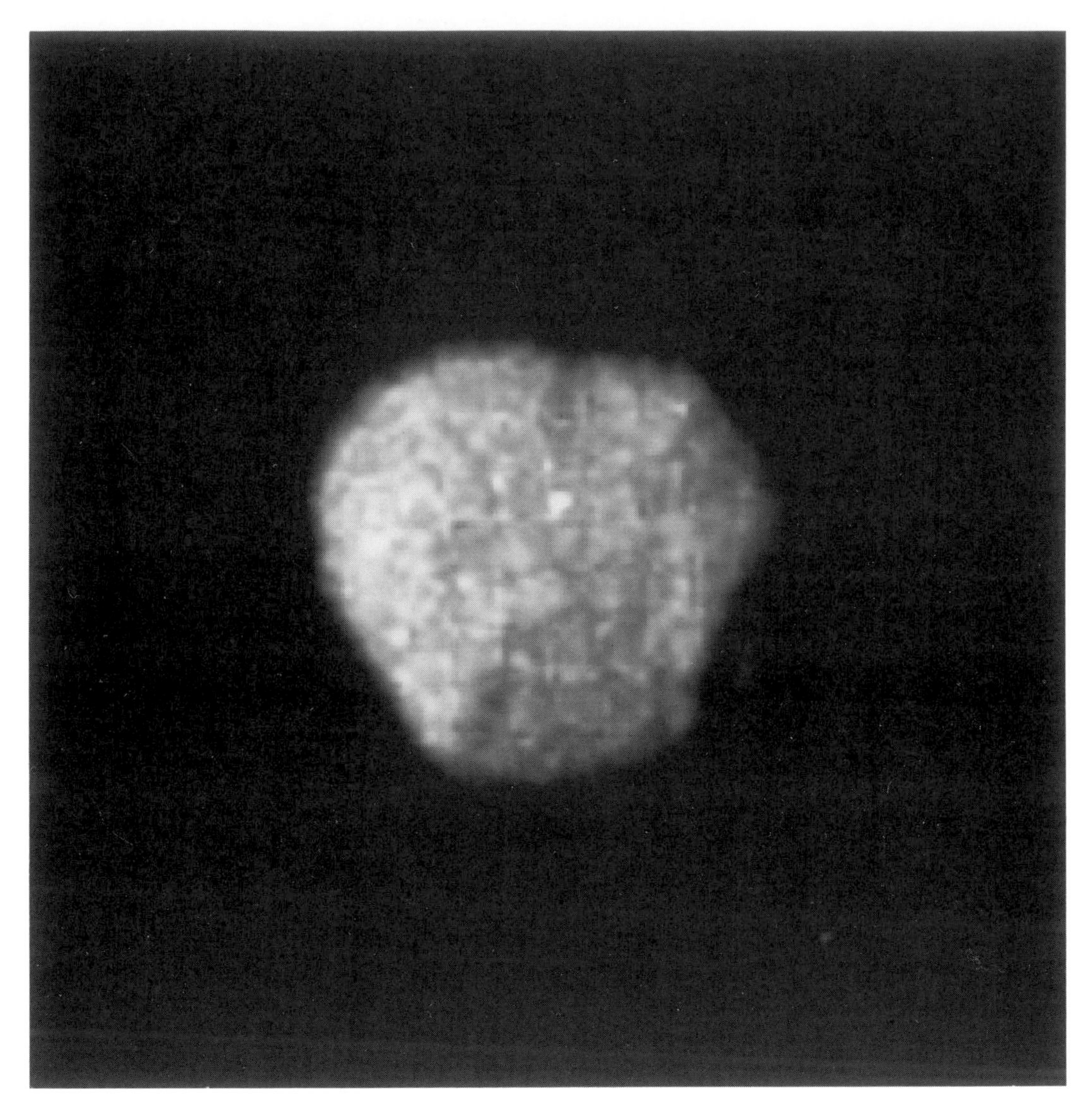

FIGURE 5.12 (B) Taken at closer range this image of 1989N1 shows more detail of the jumbled dark surface. The irregular shape of this satellite suggests that it has been cold and rigid throughout its history and has been subjected to a significant amount of impact cratering. (NASA/JPL)

As the spacecraft moved ever closer to Neptune but was still 22 million miles (35.4 million km) from the planet, three additional satellites were discovered. These three were tentatively named 1989N2 (figure 5.13), 1989N3, and 1989N4 in order of their discovery. They orbited in the region where the ring arcs were expected. These satellites were smaller than 1989N1, but they, too, orbited on the equatorial plane in orbits close to circles. All the satellites discovered by *Voyager* revolved around Neptune in the same direction as the planet's rotation, making Triton more definitely an oddball satellite since it revolves in the opposite direction. In order of distance from Neptune the three new satellites have orbits with radii of 32,620 miles (52,500 km) for 1989N3, 38,530 miles (62,000 km) for 1989N4, and 45,730 miles (73,600 km) for 1989N2.

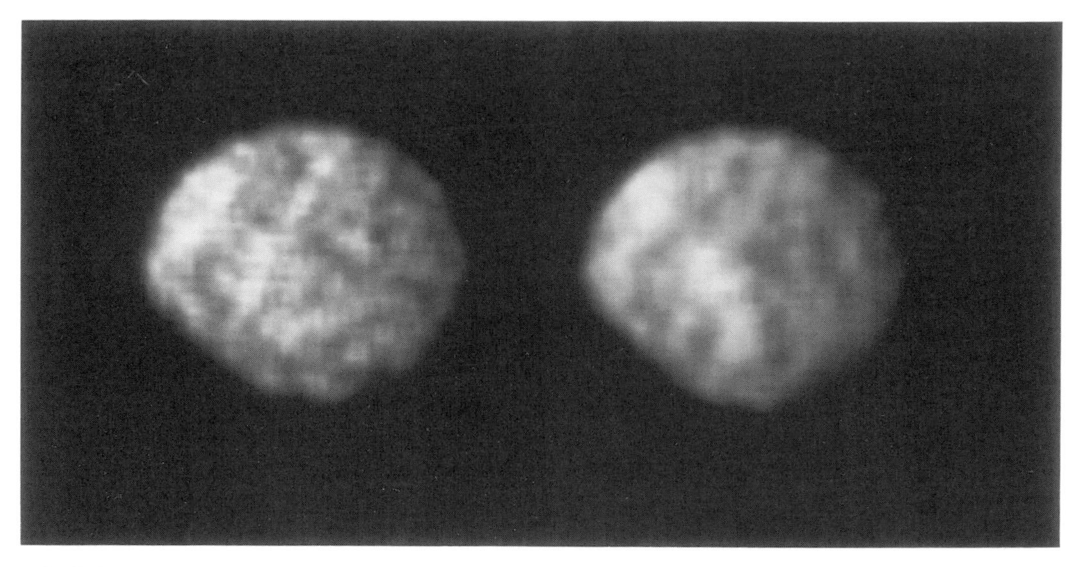

FIGURE 5.13 The *Voyager* images of the satellite 1989N2 reveal that it, too, is an irregularly-shaped dark object. There are several large craters of as much as 30 miles (50 km) diameter on the 118-mile (190-km) diameter world almost big enough to shatter it. Again, the irregular shape suggests that it has remained cold and rigid throughout its history. Its surface reflects about five percent of incident sunlight, similar to 1989N1. (NASA/JPL)

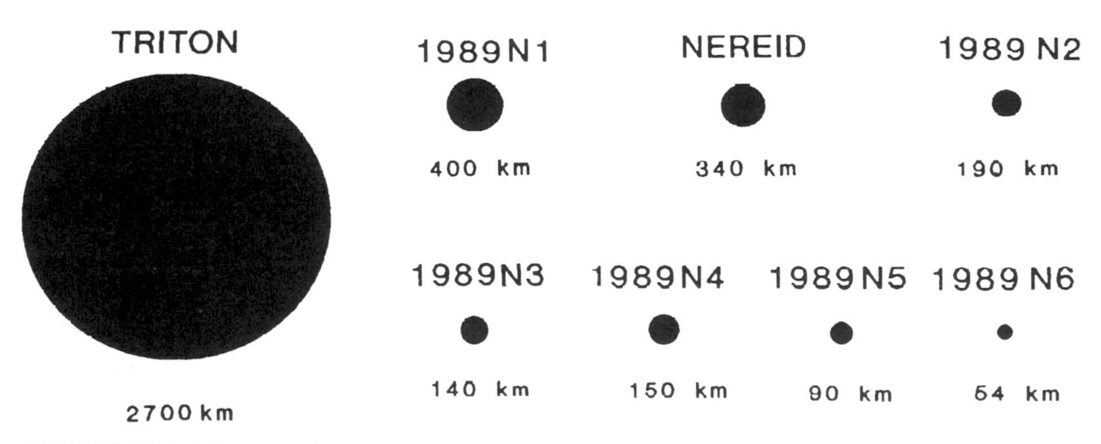

FIGURE 5.14 The sizes of the various satellites of Neptune as determined by the *Voyager* encounter are shown in this drawing.

Their respective orbital periods were established as 8 hours 10 minutes, 9 hours 30 minutes, and 13 hours 20 minutes.

Two more satellites were discovered later; 1989N5 is about 56 miles (90 km) in diameter and it orbits Neptune in 7 hours 30 minutes at a distance of 31,700 miles (50,000 km). The final satellite to be discovered during the encounter was 1989N6. It is the smallest of the newly discovered satellites. Its diameter is 34 miles (54 km) and it orbits Neptune in 7 hours 6 minutes in a circular orbit with a radius of 29,950 miles (48,200 km). These two satellites also orbit in the equatorial plane in the same direction as the planet rotates. All seemed irregularly shaped and all seemed to possess cratered

TABLE 5.2 Satellites of Neptune

Name	*Orbital Radius* *miles*	*km*	*Orbital Period*	*Orbital Inclination*	*Diameter* *miles*	*km*
1989N6	29,950	48,200	7.1 hr	4.5	34	54
1989N5	31,070	50,000	7.5 hr	<1.0	56	90
1989N4	38,530	62,000	9.5 hr	<1.0	93	150
1989N3	32,620	52,500	8.0 hr	<1.0	87	140
1989N2	45,730	73,600	13.3 hr	<1.0	118	190
1989N1	73,070	117,600	26.9 hr	<1.0	249	400
Triton	220,470	354,800	141.0 hr	158.5	1,680	2,705
Nereid	342,380	551,000	8643.1 hr	27.5	210	340

surfaces. Table 5.2 provides details of the satellites. Their sizes are compared in figure 5.14 on p. 109.

The satellite 1989N1 is one of the darkest objects in the Solar System, reflecting only 6 percent of the sunlight striking its surface. Accurate measurements of images obtained during the encounter gave a diameter of 249 miles (400 km), which is larger than the previously known small satellite, Nereid. The reason 1989N1 had not been discovered by observations from Earth is that it is so close to Neptune. Its feeble light is lost in the glare of reflected sunlight whereas the mean distance of Nereid from Neptune is five times that of 1989N1 and it escapes most of the obscuration from the planet's glare.

Constraints on the flyby path required both Sun and Earth occultations at Neptune and at Triton at a time when the spacecraft would be at a high elevation angle as observed from Canberra and would be at this high elevation for at least seven hours. Also, the spacecraft should fly as close as possible to Triton. These constraints made it impossible to time the encounter for a close approach to Nereid. The best images of Nereid had to be obtained when *Voyager* was 2.9 million miles (4.7 million km) from the satellite. They showed that its surface reflects about 14 percent of incident sunlight; slightly more reflective than Earth's Moon, and twice as reflective as 1989N1. But this albedo was approximate because of the small size of the image. Nereid was viewed in a crescent phase showing only part of its surface illuminated at this best resolution image, so it was not feasible to determine its shape. Also, the rotation period could not be determined from the images nor were any surface features clearly seen or significant brightness variations that might be attributed to rotation. The image had about the same definition as the Pioneer images of the Galilean satellites of Jupiter. The low albedo of Nereid and lack of color suggests that its surface has been churned up and reworked by impacts.

By contrast, good images were obtained of 1989N1 and 1989N2; they showed that these satellites are irregularly shaped and have cratered surfaces. While albedo features were not apparent there were large surface topographical features including a large crater on one satellite, some suggestions of lineaments, and prominent limb features. On 1989N1, for example, a gigantic crater (figure 5.15) with a diameter of 90 miles (150 km) makes a huge circular depression on the terminator of the 250-mile (400-km) diameter satellite. The surface of 1989N2 appears reddish and suggests a composition similar to the surfaces of carbonaceous chondrites.

FIGURE 5.15 This image of 1989N1 was obtained from a range of 91,000 miles (146,000 km). The satellite is half illuminated. There are hints of many craters and grove-like lineaments. What appears to be a huge crater with a diameter of 90 miles (150 km) mars the top part of the terminator region. (NASA/JPL)

The inner small satellites of Neptune were probably formed from the disruption of a larger body after the major bombardment early in the history of the Solar System. There is a sufficient population of comets associated with Neptune for that type of collision to occur. Satellites also probably "seed" the rings as the result of impact by meteorites. Five of the satellites are near enough to Neptune to form rings if disrupted by comets. The larger of the satellites exhibit craters of sufficient size that if a body of similar size to that which formed the craters had collided with a smaller satellite that satellite would have been completely disrupted. If the existing small satellites of Neptune had existed at the time of the intense bombardment that cratered the big satellites they would almost certainly have been disrupted. These smaller satellites thus appear to be the remnants of larger satellites that were broken up during the tail end of the intense bombardment.

6

TRITON

In January 1987 *Voyager* obtained its first image of Neptune's strange large satellite Triton. Neptune was still 850 million miles (1.37 billion km) ahead of the speeding spacecraft. The resolution of the image was about the same as can be obtained with the best telescopes from Earth. The purpose of these early images was to aid the navigation team to target precisely the encounter with Neptune and Triton. Because Triton's orbit is highly inclined to Neptune's equator the path of the spacecraft had to be bent by Neptune's gravity so that *Voyager* could pass close to the large satellite. At the time of the flyby Triton would be below the ecliptic plane and the spacecraft would have to pass close to Neptune over a high northern latitude of the planet to obtain sufficient bending of the spacecraft's trajectory. Precise timing and knowledge of the positions of Neptune, Triton, and the spacecraft were mandatory for the success of this maneuver in which the spacecraft would skim a mere 3000 miles (4850 km) over the cloud tops of Neptune some 3 billion miles (4.5 billion km) from Earth.

At 426 million miles (685 million km) distance in May 1988 Triton was still a mere dot on the returned images, occupying one pixel only of the image frame that has 640,000 pixels on an 800 by 800 grid. But toward the end of August images (figure 6.1) started to reveal surface features that rotated around the satellite at about the same rate as Triton's 5.88-day orbital period, indicating that the satellite did turn the same face to Neptune as Earth's Moon turns one face toward Earth.

FIGURE 6.1 Already intriguing patterns of unknown origin began to appear on images returned from *Voyager* on its approach to Neptune. This image was taken August 22, 1989 from a distance of 2.5 million miles (4 million km). This is the hemisphere of Triton that always faces away from Neptune, and the south pole of the satellite is near the bottom of the image. Dark regions at the top of the satellite's disk extend from the equatorial region to about 20 degrees north latitude. (NASA/JPL)

As the spacecraft drew even closer to the Neptunian system estimates of the size of Triton began to shrink. Said Bradford Smith, imaging team leader, "It will surely turn out to be between 1380 and 1400 kilometers radius."

Early on the morning of August 21, 1989 the spacecraft performed a trajectory correction maneuver. Norman Haynes, project manager, explained that the maneuver was designed to trim *Voyager*'s trajectory past Triton. "We have no safety concerns," he said. "Before the TCM (trajectory correction maneuver), *Voyager* would have performed the Earth occultation, but not the Sun occultation. Now we'll do both."

The maneuver altered *Voyager*'s speed by about 0.5 meters per second to change the flight path's direction. The navigation team's outstanding performance had placed the spacecraft on the right trajectory for a successful encounter with Triton.

By the next day (August 22) images in color were showing that Triton had a mottled surface in a variety of pinkish hues. What these surface features were would not be revealed until *Voyager* passed its closest approach to Neptune and could start to concentrate its observations on the big satellite. But the view of Triton even now began to look very intriguing.

On August 23 an image returned from the spacecraft (figure 6.2) at a distance of 1.7

million miles (2.7 million km) showed features about 31 miles (50 km) across. Bright wisps extended northeast from the boundary between light and dark regions that coincided with the satellite's equator.

Arrangements were soon being made to correct the trajectory of the spacecraft for the Triton flyby. On August 23, the project manager, Norman Haynes, showed how the current trajectory of *Voyager* differed from the desired aim point to obtain an occultation of both Sun and Earth as the spacecraft would fly behind Triton (figure 6.3). The position of the desired aiming point at Triton is shown in the figure. At the enormous distance of the spacecraft from Earth its path had to be such as to pass by Triton at a distance of 24,500 miles (39,500 km) through an occultation zone that was only just over 680 miles (1000 km) wide. The uncertainty in the position of the spacecraft within the occultation zone was about 150 miles (240 km) so the aim point had to be centered within the occultation zone as closely as possible. In addition, the timing had to be exact so that commands to the spacecraft would be executed at the correct time for ingress and egress radio experiments and for directing the radio beams to reach Earth despite bending of the beams by the atmosphere of Triton.

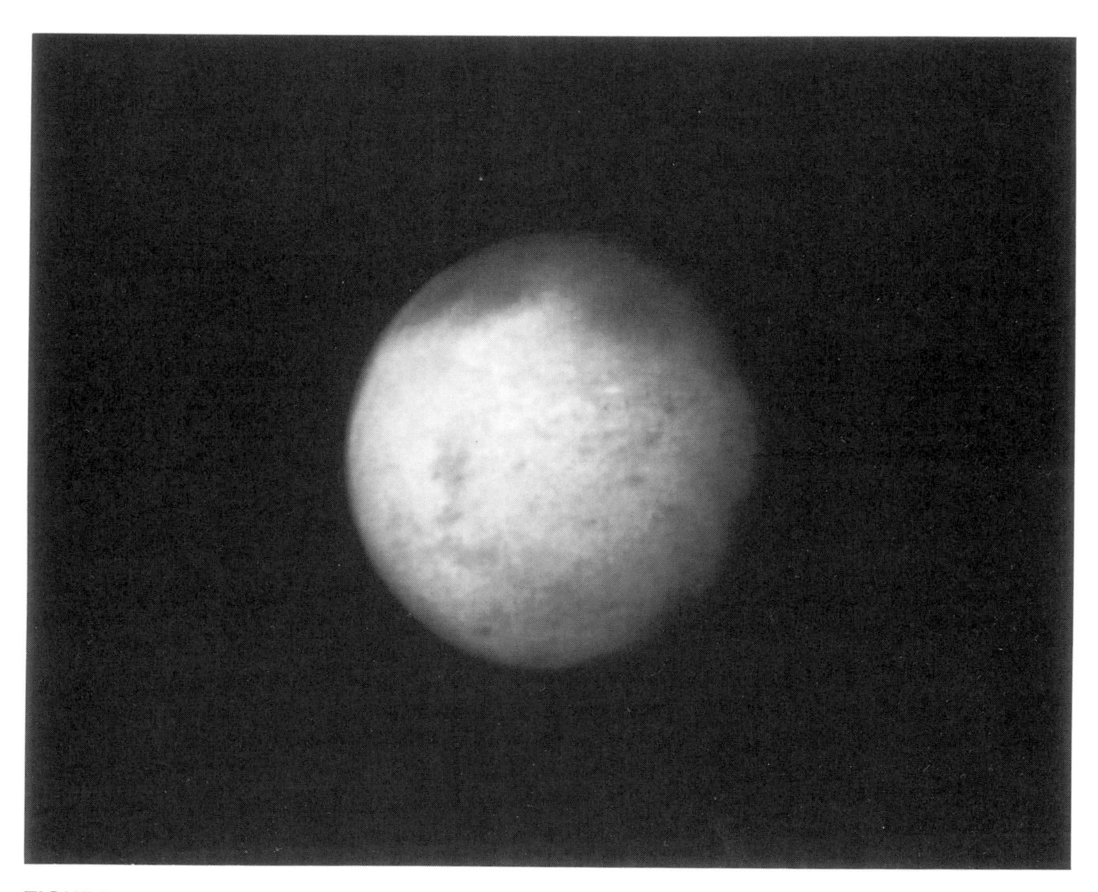

FIGURE 6.2 At 1.7 million miles (2.7 million km) Triton begins to reveal more details of its bizarre surface, but still the smallest feature that can be seen on the images is about 30 miles (50 km) across. The boundary between the dark and light regions in the upper part of the image nearly coincides with Triton's equator. Bright wisps extend northeastern from the boundary. They may be frost deposits condensing from atmospheric gases moving toward the cold northern regions. (NASA/JPL)

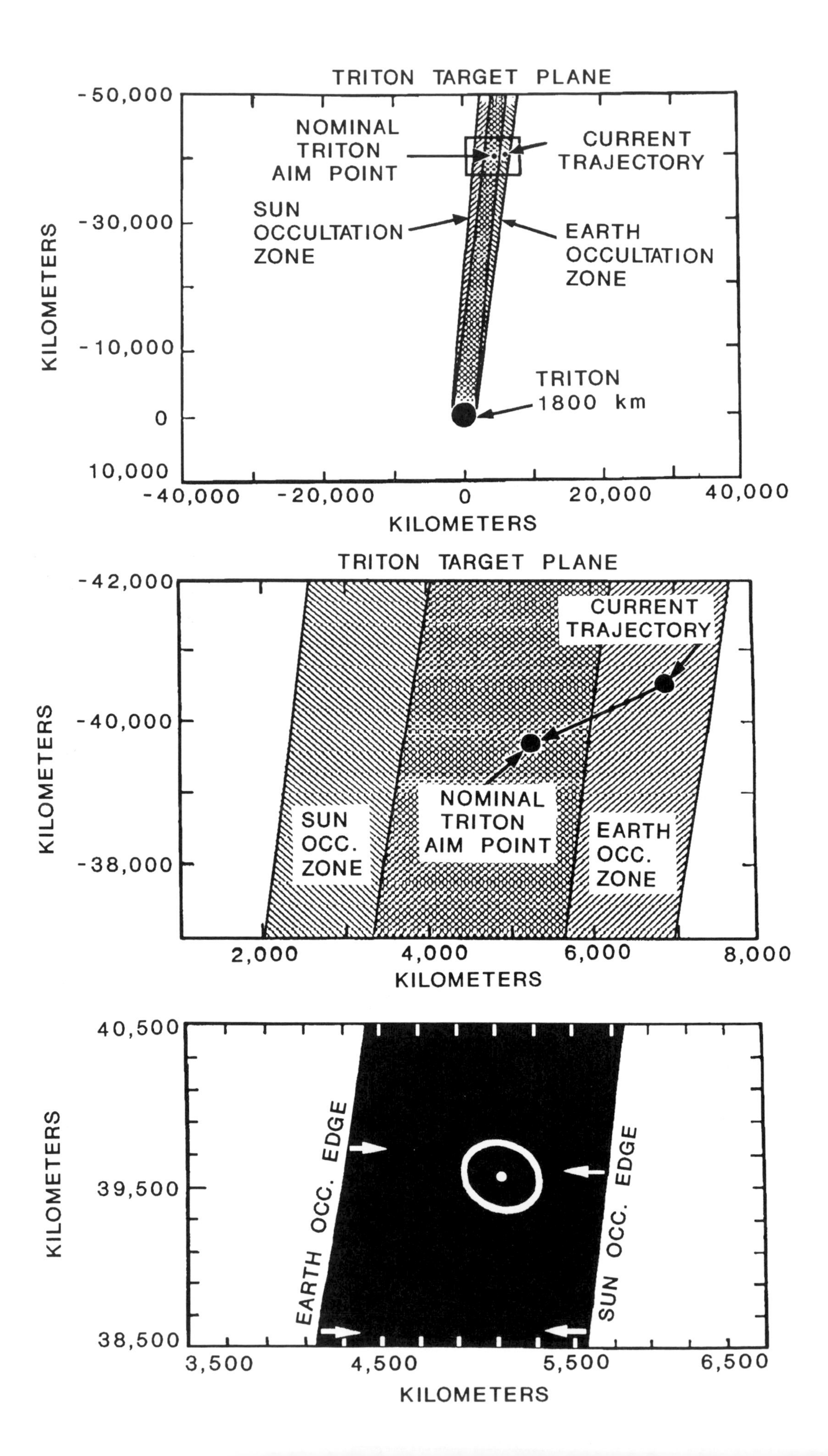
TRITON TARGET PLANE
-50,000
-30,000
-10,000
0
10,000
KILOMETERS
NOMINAL
TRITON
AIM POINT
CURRENT
TRAJECTORY
SUN
OCCULTATION
ZONE
EARTH
OCCULTATION
ZONE
TRITON
1800 km
-40,000
-20,000
0
20,000
40,000
KILOMETERS
TRITON TARGET PLANE
-42,000
-40,000
-38,000
KILOMETERS
CURRENT
TRAJECTORY
SUN
OCC.
ZONE
NOMINAL
TRITON
AIM POINT
EARTH
OCC.
ZONE
2,000
4,000
6,000
8,000
KILOMETERS
40,500
39,500
38,500
KILOMETERS
EARTH OCC. EDGE
SUN OCC. EDGE
3,500
4,500
5,500
6,500
KILOMETERS

Transmissions containing the commands to be loaded into the spacecraft's memory for the encounter sequence were sent six times to be sure they were received. All were received. The commands were adjusted slightly on August 24. Haynes reported that *Voyager 2* was "right where we want it." By early morning on August 25 the close encounters of Neptune and Triton had been completed. "We have seen Voyager's last planetary target" said imaging team member Larry Soderblom immediately after the encounters. "All we can say is, 'Wow! What a way to leave the Solar System.' "

Closest approach to Triton took place early in the morning of August 24, just over five hours after closest approach to Neptune. Detailed imaging of the satellite began soon after the spacecraft emerged from occultation by Neptune. After an overnight watch by many engineers, scientists, media representatives, and others at the Jet Propulsion Laboratory, excitement grew among the watchers as images of greater and greater detail were displayed. In the press room at the Von Karman auditorium groups of reporters gathered around the TV monitors, intently watching the fascinating details of the bizarre world that were being revealed to human eyes for the first time.

A bluish and pinkish world, Triton displayed an unusual and broad mixture of terrains. Some features resembled those on many other planets and satellites, while others had never been seen before on any planetary body. There is now little doubt that the color of Triton has been changing over the last decade and losing some of its redness. Today, as seen from *Voyager* and determined by use of various color filters, Triton has a slightly reddish central polar region, a neutral colored bright band around the outer region of the polar cap, and more reddish regions extending into the northern hemisphere. Some reddish-tinged surface also appears to be concentrated toward the trailing hemisphere of the satellite. The reddish color probably comes from organic polymers derived from methane by the action of cosmic rays, solar radiation, and charged particles from Neptune's plasmasphere. The bright neutral collar is probably freshly deposited, and hence unmodified, ices.

Triton has a remarkably high albedo. Its surface reflects a very large proportion of incident sunlight. This is consistent with a surface that is being revised on a continuing basis. This surface is more reflective than those of other icy satellites such as the satellites of Saturn. The unusual surface of Triton probably results from the cycles of sublimation and condensation that are caused by Triton's axial and orbital inclinations.

Images of the whole disk (figure 6.4) revealed three very different regions; a bright south polar region, a dark northern hemisphere, and a bright subequatorial band along the edge of the polar cap. The boundary between the bright southern hemisphere and the darker northern hemisphere was clearly displayed due to the tilt of Triton's axis and the highly inclined orbit.

At this epoch the southern hemisphere of the big satellite receives more direct sunlight and is covered with patterns of light and dark regions. The south pole is close to the middle of its 41-year-long summer. Ices are subliming and the gases from them are migrating across the equator to the north polar region. This region was, however,

FIGURE 6.3 The aiming point for the encounter with Triton was quite critical. The top picture shows Triton and the zones of Sun and Earth occultation that the spacecraft must pass through. On the next picture the nominal aim point and the current trajectory are shown in the rectangular portion enlarged from the top picture. At the bottom is the final aim point of the spacecraft within a circle of possible error. *Voyager* was right on target for the encounter: a tremendous feat of celestial navigation. (JPL)

FIGURE 6.4 This picture of Triton was assembled from 14 individual frames. At the left are the remnants of the south polar cap. It contains dark streaks aligned generally toward the northeast (lower right on the image). Even though they appear dark, these streaks reflect ten times as much light as the surface of Earth's Moon, which is a very dark body. North of the cap in the western (top) half of the disk is what has been informally dubbed the "cantaloupe" region. Running east to the limb of Triton, just north of the polar cap, is an area of relatively smooth plains and low hills which is the most densely cratered region seen on the satellite. In the northeast (lower right) of this image are plains which show evidence of extensive resurfacing, and contain two large smooth areas similar to lunar maria. (NASA/JPL)

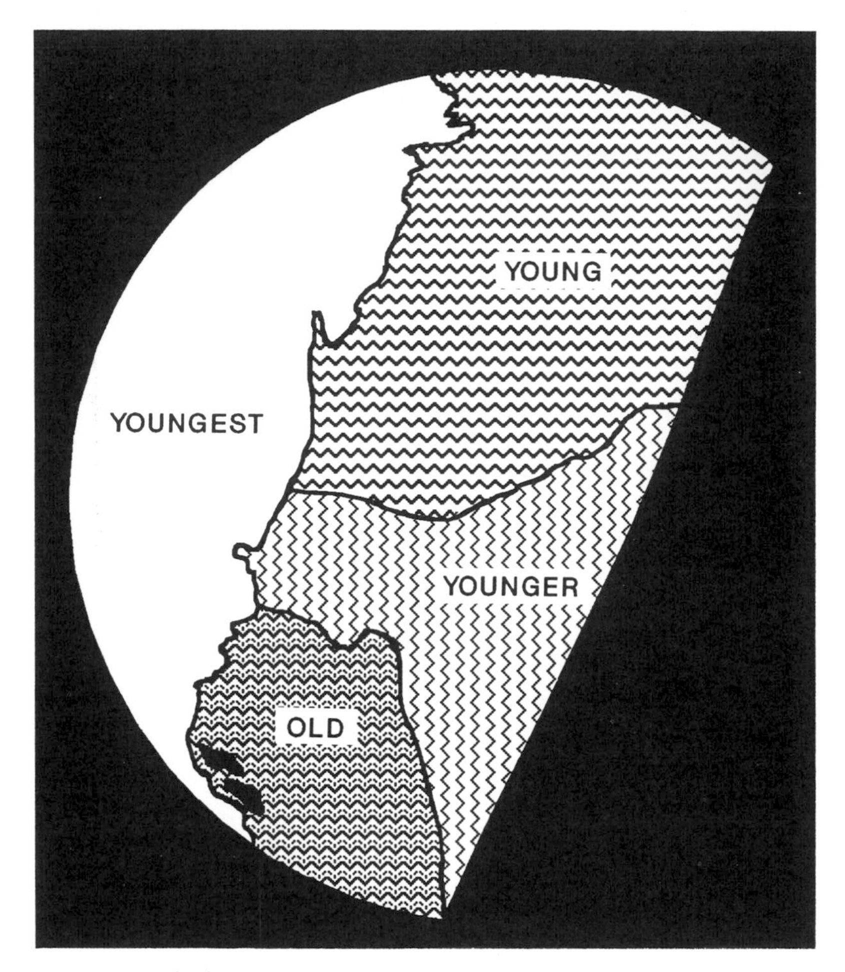

FIGURE 6.5 Estimates of relative ages for the different areas of Triton's surface imaged in Figure 6.4 are shown on this drawing. These ages were suggested by some members of the imaging team.

hidden from the cameras of the spacecraft on the hemisphere of Triton in darkness at the time of the flyby. There are also long straight lines that are believed to show the presence of internal tectonic processes, and some evidence of strike-slip faulting where surface features are displaced laterally.

Since there are no large impact craters and very few impact craters of any size up to the limits of resolution of the images, the crust of Triton must have been renewed comparatively recently—within the past billion years. Exactly how recently is not known. But in general Triton's surface appears comparatively young; only millions to hundreds of millions of years old compared with billions of years for bodies such as Earth's Moon and many other satellites in the Solar System. Relative ages of different areas shown on figure 6.4 are suggested in figure 6.5. Topographic features, many as high as 3000 feet (1 km) seen on the surface of Triton are believed to consist of water ice because at the temperature of Triton such ice behaves like rock on the terrestrial planets. Methane and nitrogen ices do not have the strength to support their own mass so that features consisting of such ices would deform fairly rapidly.

Methane and nitrogen ices probably exist as a veneer over the underlying water ices

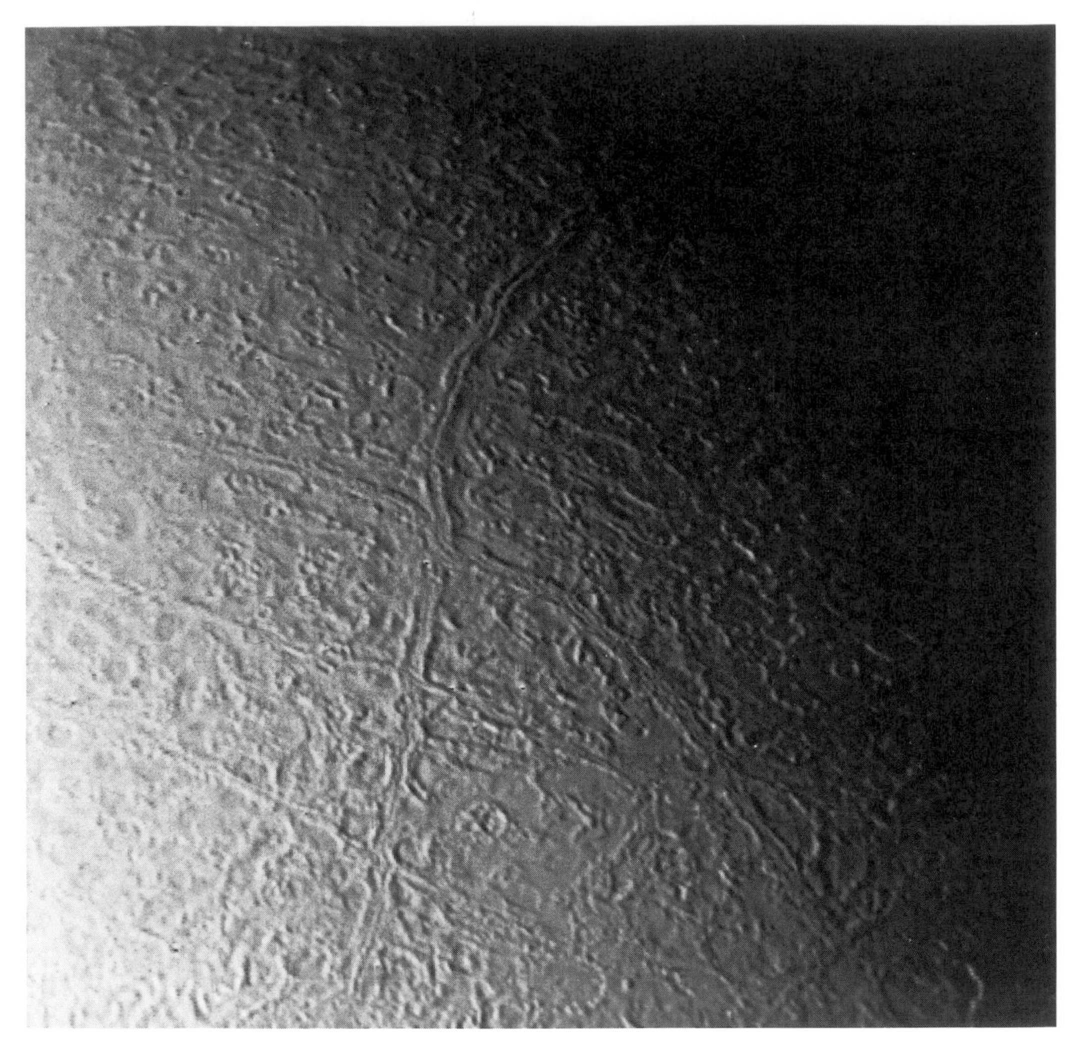

FIGURE 6.6 This *Voyager* image obtained during close approach to Triton show details of the "cantaloupe" terrain. A complex surface texture is revealed on which there are some circular features that are probably not impact craters. Strange lineaments crisscross the area. (NASA/JPL)

which account for the topographic features. Most albedo differences result from this veneer and do not appear to be associated with topographical features. Triton is large enough for it to have contained sufficient radioactive material to cause its materials to differentiate after accretion. The result would be a core of silicates surrounded by a mantle of ices several hundred kilometers in thickness. The surface features appear to be consistent with the theory that the big satellite suffered tidal heating and extensive resurfacing also, perhaps for as long as a billion years. In turn this is consistent with the speculation that Triton was captured by Neptune and is a planetary body similar to Pluto. Even today tidal forces on this retrograde satellite are pulling it closer and closer to Neptune, and probably still developing heat within the satellite.

In Triton's northern hemisphere there are large tracts of peculiar terrain. A vast area has roughly circular depressions separated by rugged ridges (figure 6.6). These features may result from local melting and collapse of the icy surface.

The surface of Triton appears as though, comparatively recently, some regions have been soft while others have remained rigid, which would support the viewpoint of continued internal heating. Soderblom pointed out at a press conference that "We see bowl-shaped craters about 20 kilometers [12 miles] across and two kilometers [1.2 miles] deep where the surface must have been rigid, and places where it seems to have been soft." Some pictures from *Voyager* showed features that were similar to volcanic calderas on rocky planets such as Venus and Mars. Two great depressions near to the terminator (figures 6.7 and 6.8) were possibly impact basins that were subsequently modified greatly by flooding, melting, faulting, and collapse. Material had apparently risen and collapsed within the flat areas at different levels, just as lava rises and collapses to different levels on terrestrial volcanoes such as Kilaeua caldera on the Big Island of

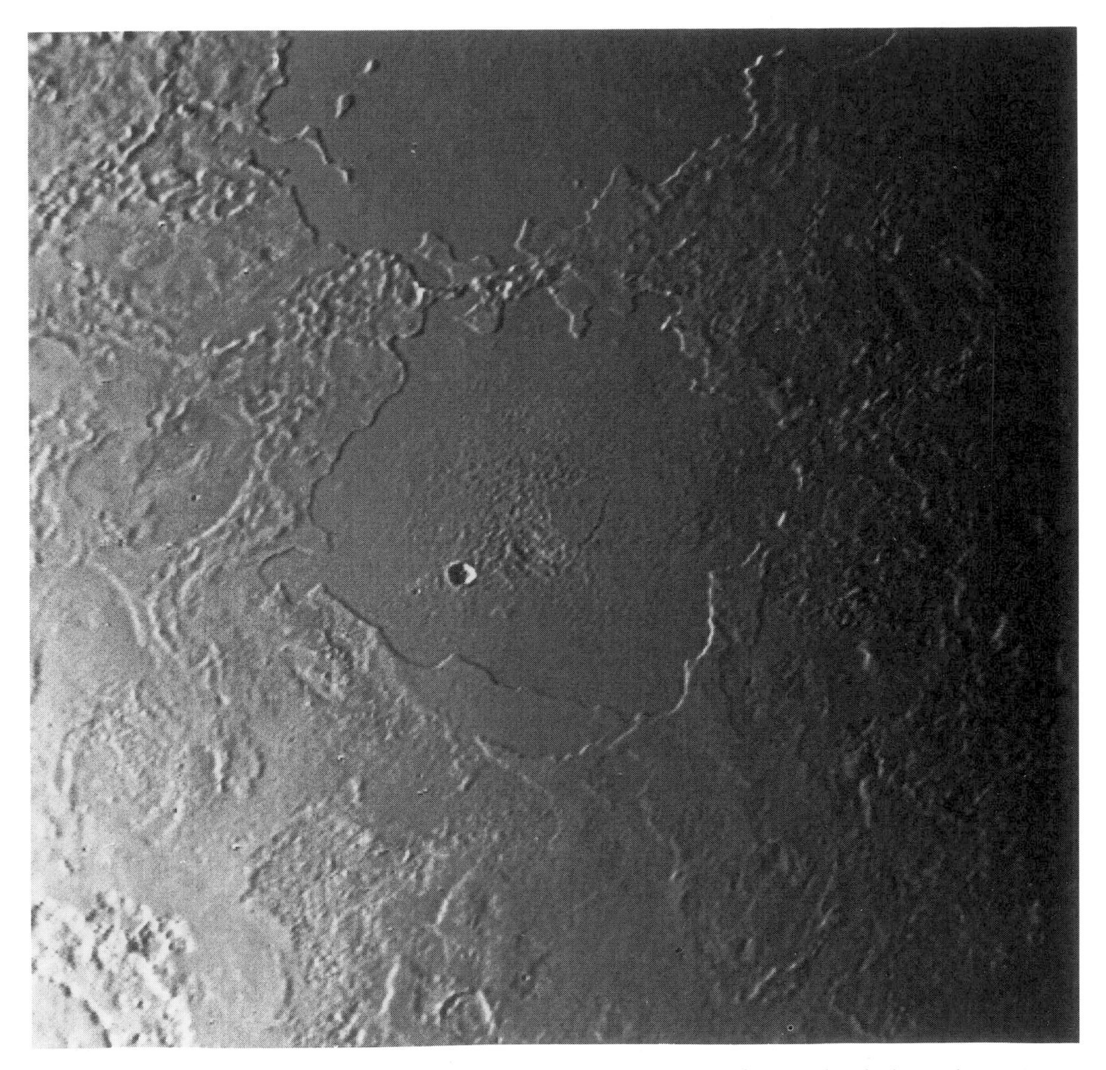

FIGURE 6.7 Part of the complex geological history of Triton is evident in this high resolution image which is about 300 miles (500 km) across. Two extensive depressions, possibly ancient impact basins, have been extensively modified by flooding, melting, faulting, and collapse. The rough area in the lower depression probably marks the most recent eruption of material into the basin. Only a few impact craters dot the area which shows the dominance of internally driven geological processes on Triton. (NASA/JPL)

FIGURE 6.8 This image of the other flooded depression suggests a formation from two large impact basins that overlapped. The basin is 120 miles (200 km) by 240 miles (400 km). A vent from which the most recent flooding occurred appears to lie near the righthand edge of the basin. The lack of impact craters suggests that this is a relatively recent episode of flooding. (NASA/JPL)

FIGURE 6.9 This image shows an area of Triton's northern hemisphere. The Sun is near the horizon so that features are thrown into relief by long shadows. The large smooth area, part of the flooded depression shown in figure 6.7, has a single impact crater. Otherwise there is no evidence of impacts that mar the surfaces of so many other satellites of the outer planets. Many low cliffs in the area, bright where they face the Sun, suggest an intricate history. These cliffs may be due to melting of surface materials or may be caused by unusual fluid materials that flowed sometime in Triton's past. (NASA/JPL)

Hawaii. The difference is that the temperature of lava that erupts on terrestrial volcanoes is more than 1000 degrees, whereas the temperature on the surface of Triton is several hundred degrees below zero Fahrenheit.

The image shown in figure 6.9 shows part of the northern hemisphere. With the Sun shining from just above the horizon, shadows are cast and features are emphasized. The large volcanic plain, which is also shown on figure 6.7, has a single, fresh impact crater, one of very few such craters on Triton. Nearby hummocks probably represent residual material associated with vents through which the depression was filled by volcanics.

Volcanic systems on Earth and some satellites of the outer Solar System are compared in table 6.1.

Another image (figure 6.10) showed features near one part of the edge of the polar cap that scientists did not immediately understand. These consisted of flat, dark regions surrounded by bright halos, but exhibiting no topography. During a press conference Soderblom offered a tentative explanation that they were evidence of a thin layer of

FIGURE 6.10 Triton's limb cuts obliquely across the middle of this image. The field of view is about 600 miles (1000 km) across. Three irregular dark areas unlike anything seen elsewhere in the Solar System dominate the image. (NASA/JPL)

TABLE 6.1 Comparison of Several Volcanic Systems

		Temperature Ranges	
	Materials	*K*	*Deg. F*
Earth	Silicates and water	400 to 1600	261 to 2421
Io	Silicates, sulfur, sulfur dioxide	200 to 1600	−99 to 2421
Ariel	Ammonia and water	173 to 273	−148 to −32
Triton	Ammonia, water, methane, nitrogen	60 to 273	−351 to −32

dark material beneath a brighter surface layer. If the surface layer were punctured the dark material beneath would absorb more heat and increase in temperature. This would cause a spreading area of surface material to boil off, thus revealing more dark material. The gas that boiled away would condense in a veneer around the spreading center to produce the light halo. Most features that look like craters on Triton are almost certainly not impact craters, commented Bradford Smith. He also reported that at this time as a result of the encounter the radius of the satellite was believed to be close to 845 miles (1360 km).

The bright polar cap covers almost the whole of the southern hemisphere (figure 6.11). The icy deposit is reddish and there are irregular patches where apparently the brighter deposit has been removed to reveal darker surface beneath. That the icy deposits are thin is evident from a lack of relief along the boundaries of the patches (see figure 6.11b). Within the cap area are darker streaks somewhat similar to the eolian streaks on the surface of Mars. Some streaks on Triton are 60 miles (100 km) long. An irregular dark patch at one end of a streak suggests an emitting vent from which the dark material comes into the atmosphere to be carried across the surface by winds.

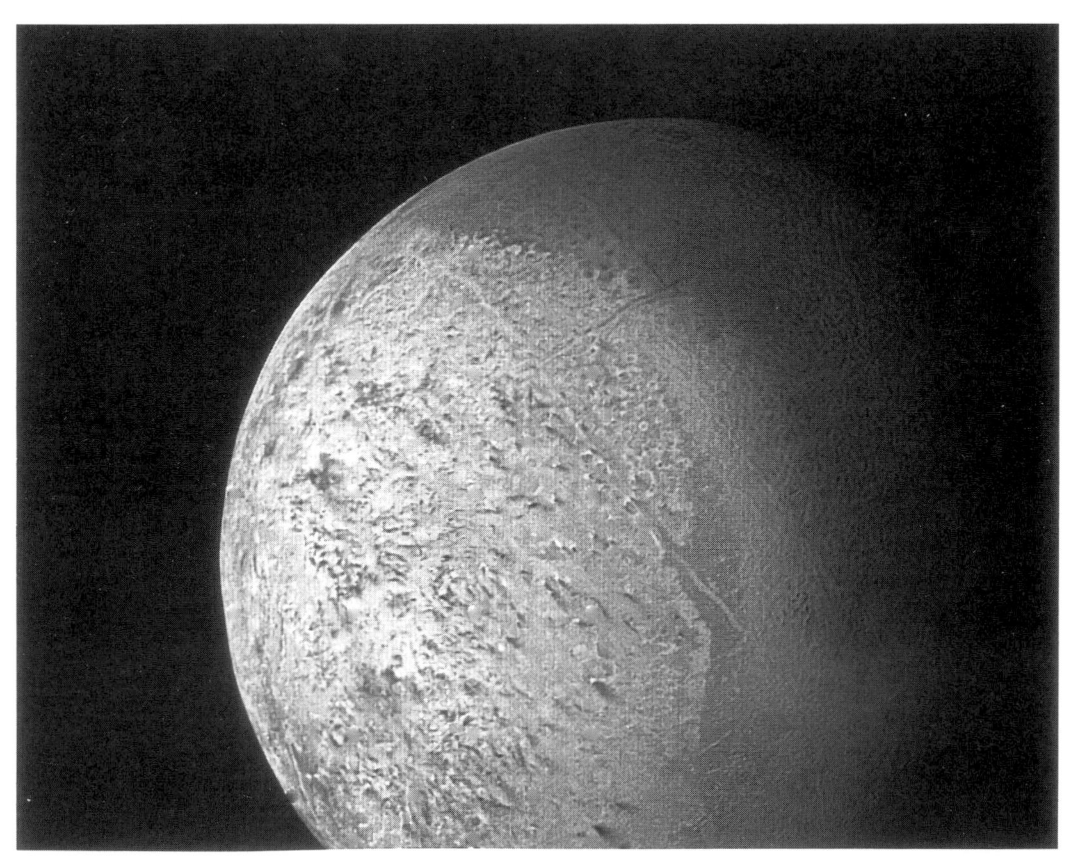

FIGURE 6.11 (A) The south pole, continuously illuminated by the Sun at this season, is at bottom left. The boundary between the bright southern hemisphere and the darker northern hemisphere is clearly visible. Both the darker regions to the north and the very bright subequatorial band show a complex pattern of irregular topography. Also evident are long straight lines that appear to be surface expressions of internal, tectonic processes. No large impact craters are visible which suggests that the surface of Triton has been renewed relatively recently, within the past billion years. (NASA/JPL)

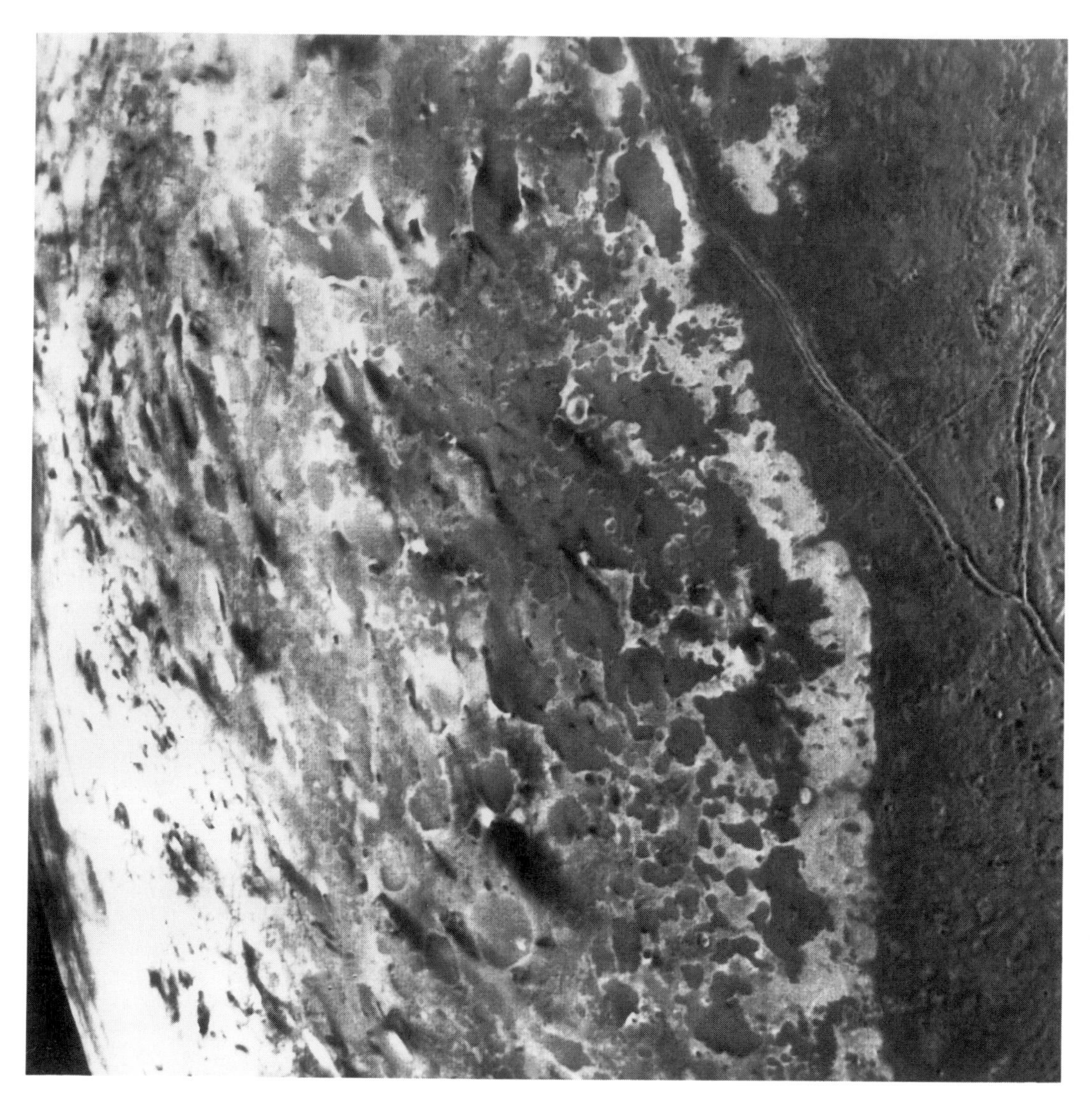

FIGURE 6.11 (B) A closeup of the south polar terrain shows dark streaks on the icy surface. The streaks originate at very dark spots generally a few miles in diameter and some of the streaks are more than 100 miles (160 km) long. The spots, which clearly mark the source of the dark material, may be vents where gas has erupted from beneath the surface and carried dark particles into Triton's nitrogen atmosphere to be carried by winds and deposited to form the streaks. The polar terrain is in general a region of bright materials mottled with irregular, somewhat darker patches. (NASA/JPL)

Close to the pole the streaks appear randomly directed, while at lower latitudes they trend toward the northeast. At present the atmospheric pressure on Triton is insufficient for surface winds to raise particles into the atmosphere to produce the dark streaks seen on the images of the polar region. The surface materials have somehow to be ejected from the surface into the atmosphere before they can be carried by a wind on Triton. The discovery of plumes of material emitted by what are presumed to be ice volcanoes suggests that these streaks are of recent origin. Indeed, it is unlikely they could be ancient because the dark material would be covered by frosts or would collect solar heat and sink through a frosty surface.

Beyond the south polar cap there are many lineaments in the complicated terrain of the northern hemisphere where patterns resemble those on the skin of a cantaloupe (figure 6.12). Material welled up on the floors of these wide valleys and formed distinctive ridges, some of which rise higher than the surrounding terrain (figure 6.13).

Another area of the northern hemisphere exhibits an entirely different type of terrain from the jumbled cantaloupe pattern. Bordering the edge of the polar cap this surface unit consists of comparatively smooth plains on which the few impact craters visible on the surface are located (figure 6.14, p. 129). At first this might be considered an older unit because of the presence of these craters. However, it occupies a position on the hemisphere of Triton that is always at the front of the satellite in its motion around Neptune. Accordingly, this leading edge would be expected to encounter more space debris than other parts of the satellite and thus have more impact craters. But this effect may not account in total for the cratering, and the unit may be an older one. The unit

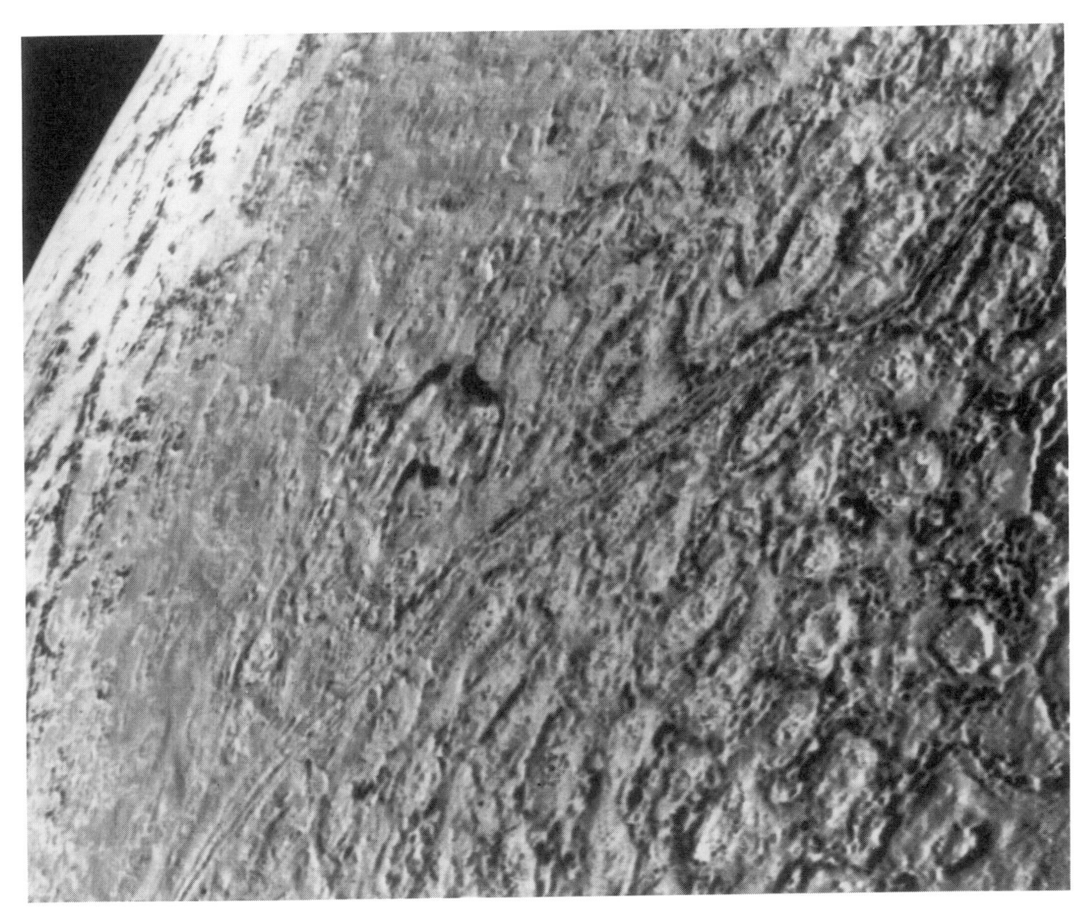

FIGURE 6.12 This is one of the most detailed views of the surface of Triton taken by the imaging system of *Voyager.* The image is about 140 miles (220 km) across and shows details as small as 800 yards (750 m) across. Most of the landscape is covered by roughly circular depressions separated by rugged ridges. This peculiar type of terrain is unlike anything seen elsewhere in the Solar System. The origin of the circular features may involve local melting and collapse of an icy surface. A conspicuous set of grooves and ridges cuts across the landscape, indicating fracturing and deformation of the surface. The rarity of impact craters suggests this is a very young surface. (NASA/JPL)

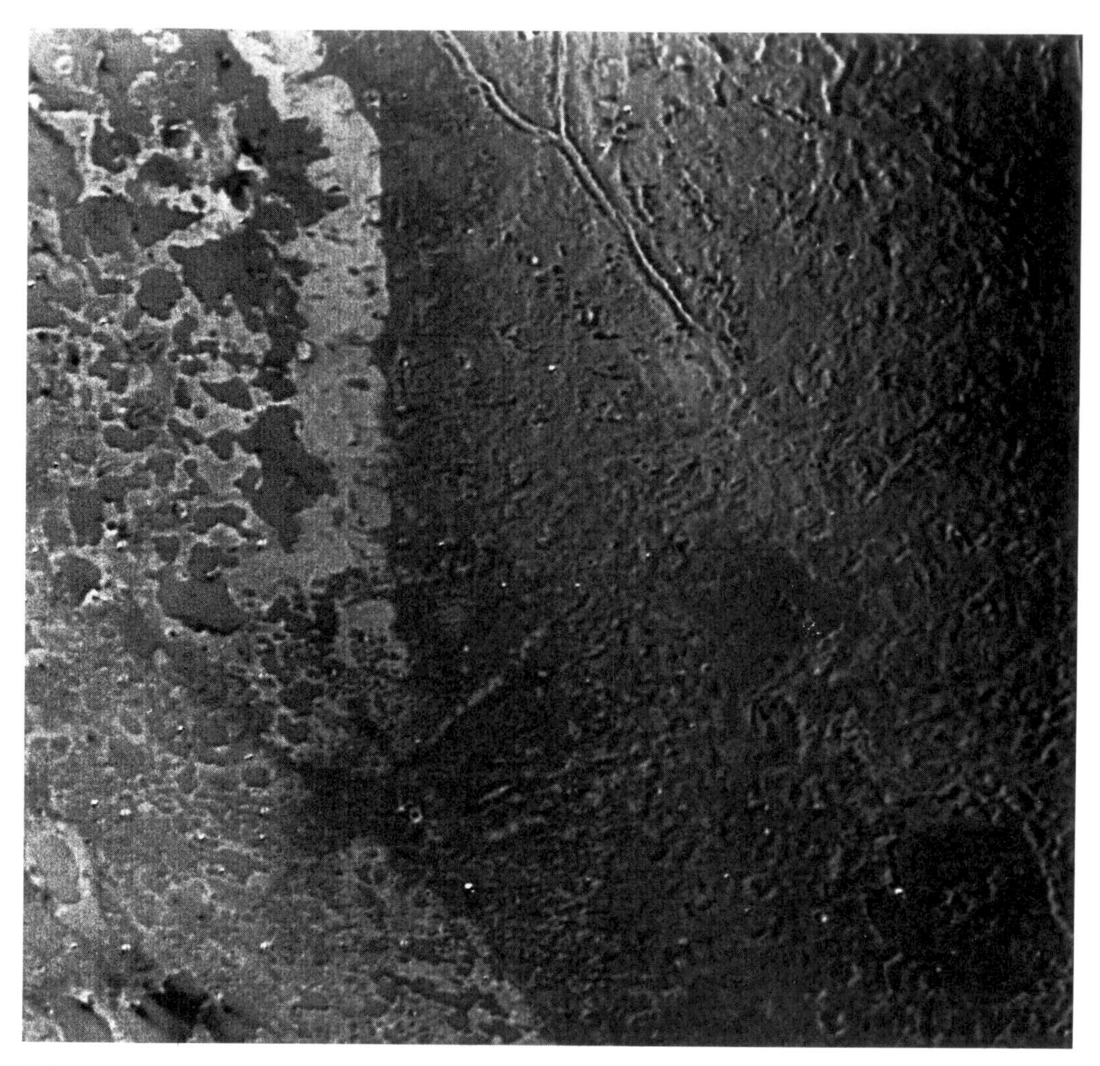

FIGURE 6.13 The surface in this image is dominated by many roughly circular, polygonal, and arcuate features between 18 and 30 miles (30 and 50 km) across. Peculiar intersecting double ridged lines are 9 to 12 miles (18 to 20 km) wide and hundreds of miles long. Patches of plains forming material occurs in depressions. (NASA/JPL)

is characterized by flows that have produced hummocky plains and were of sufficient volume to obliterate many of the grabens from which they reached the surface.

A geyser-like eruption of dark material proved that volcanic activity persists today on Triton. This volcanic activity with a 5-mile high plume and 90-mile (150-km) long cloud was discovered early October 1989 just as the encounter ended its fourth phase as scientists continued to examine in more detail many images returned by *Voyager.* The cloud stretched in a northwest direction from the plume that shot five miles (eight km) vertically from a vent on the surface of the polar cap. A distinct shadow was cast on the bright surface (figure 6.15). This is the first time that any such activity has been seen on a planetary body other than the Earth and Io. The narrow stem of the dark plume resembles a smokestack until it reaches five miles at which level it spreads out into a small dark cloud from which a long narrow cloud extends high in the satellite's atmosphere.

One possible mechanism is an artesian type eruption based on phase change without requiring much heat. Such a type of eruption was suggested for Jupiter's satellite Io in addition to the "Pele" type in which sulfur was heated by hot rocks on Io and expelled at high velocity like a rocket exhaust. The artesian type was suggested for Io to account for light-colored deposits along the bases of scarps. In an artesian-type of plume

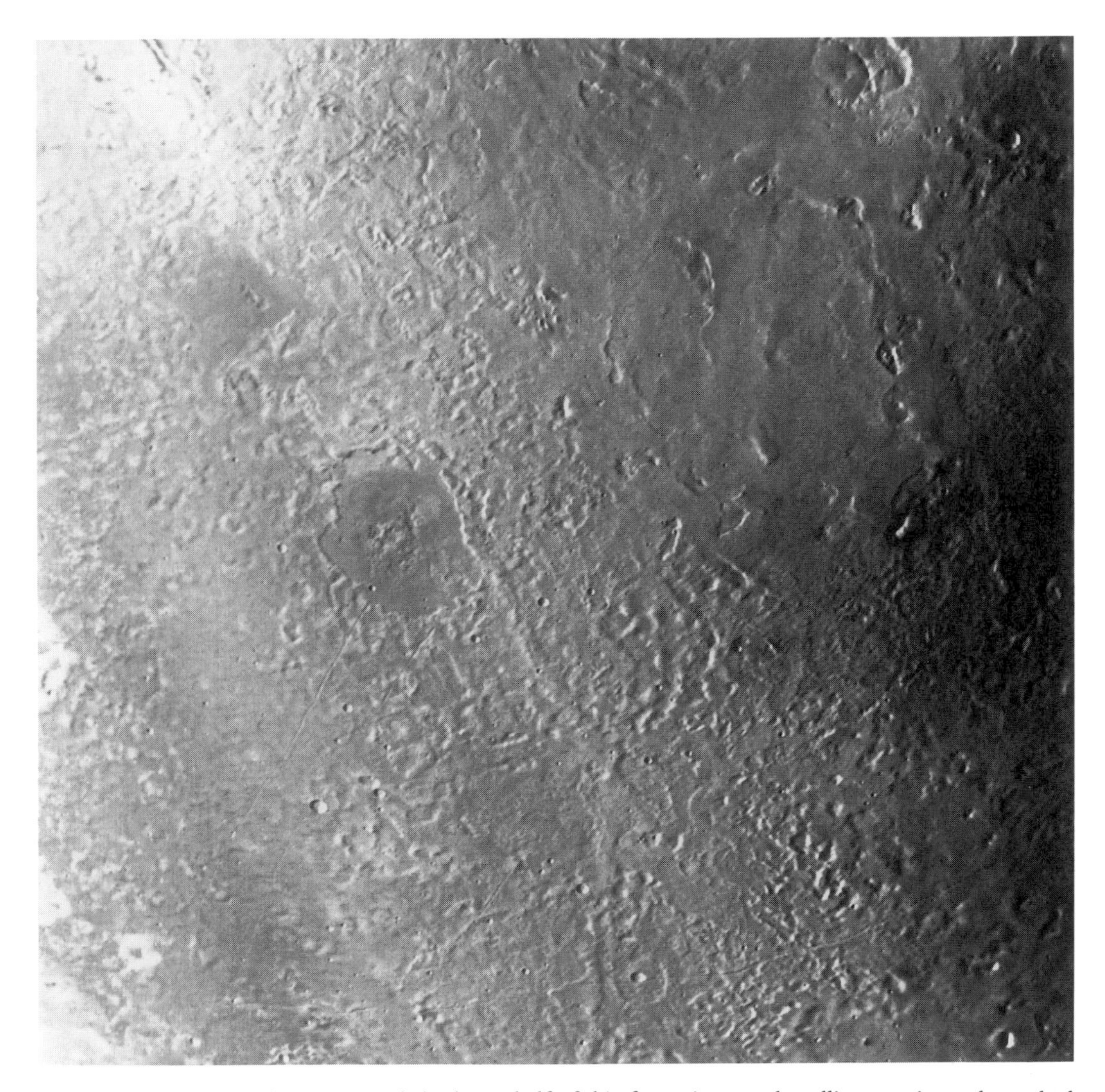

FIGURE 6.14 Near the center and the lower half of this frame is a gently rolling terrain pock-marked with a modest number of impact craters. Their density is comparable with that on the mare surfaces of our Moon. Crossing this rolling surface are narrow rifts one of which grades into a chain of craters that probably originated from a collapse of the surface. In the upper right part of the frame is an area of smooth terrain with very few impact craters. Evidently this terrain was formed by a comparatively recent flooding by low viscosity fluids. A deep, elongate crater near the middle of the right side of the image could have been the vent through which the floods emerged. (NASA/JPL)

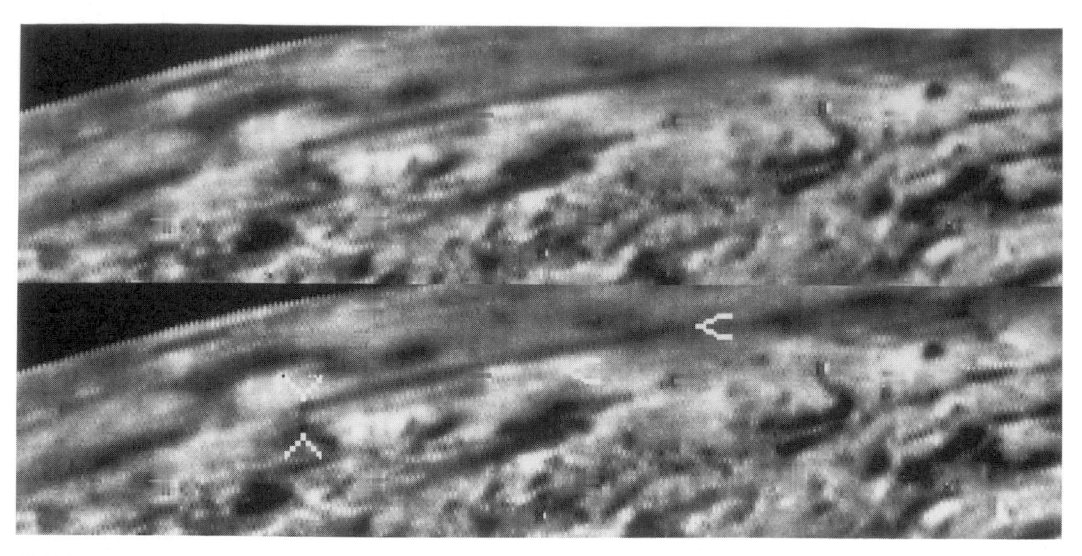

FIGURE 6.15 A tall geyser-like eruption of dark material is seen shooting almost straight up from the surface of Triton in this image looking toward the limb of the south polar region. The narrow stem rises nearly five miles (eight km) and then forms a cloud which drifts to the right some 90 miles (150 km). The lower picture is marked to show the plume and the lateral extent to the cloud. (NASA/JPL)

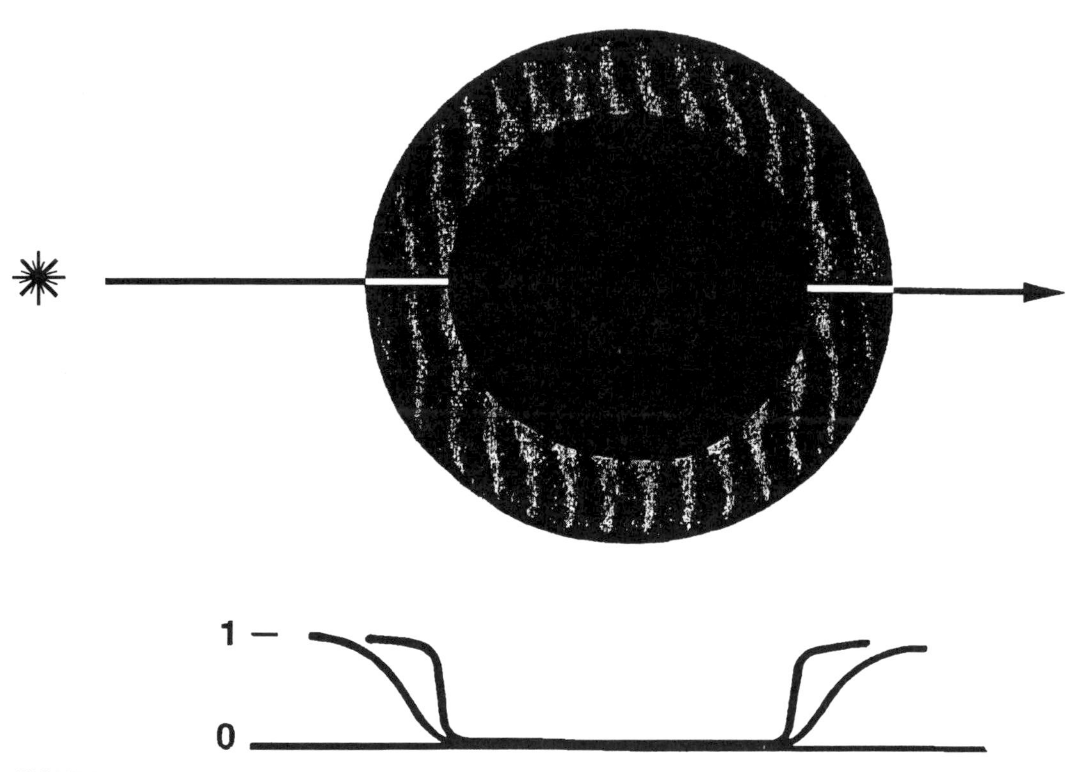

FIGURE 6.16 Diagram to show the occultation of the star Beta Canis Majoris by Triton as observed from the *Voyager* spacecraft. The curves show how the light was gradually dimmed as the starlight passed deeper and deeper through the atmosphere until it was cut off at the surface.

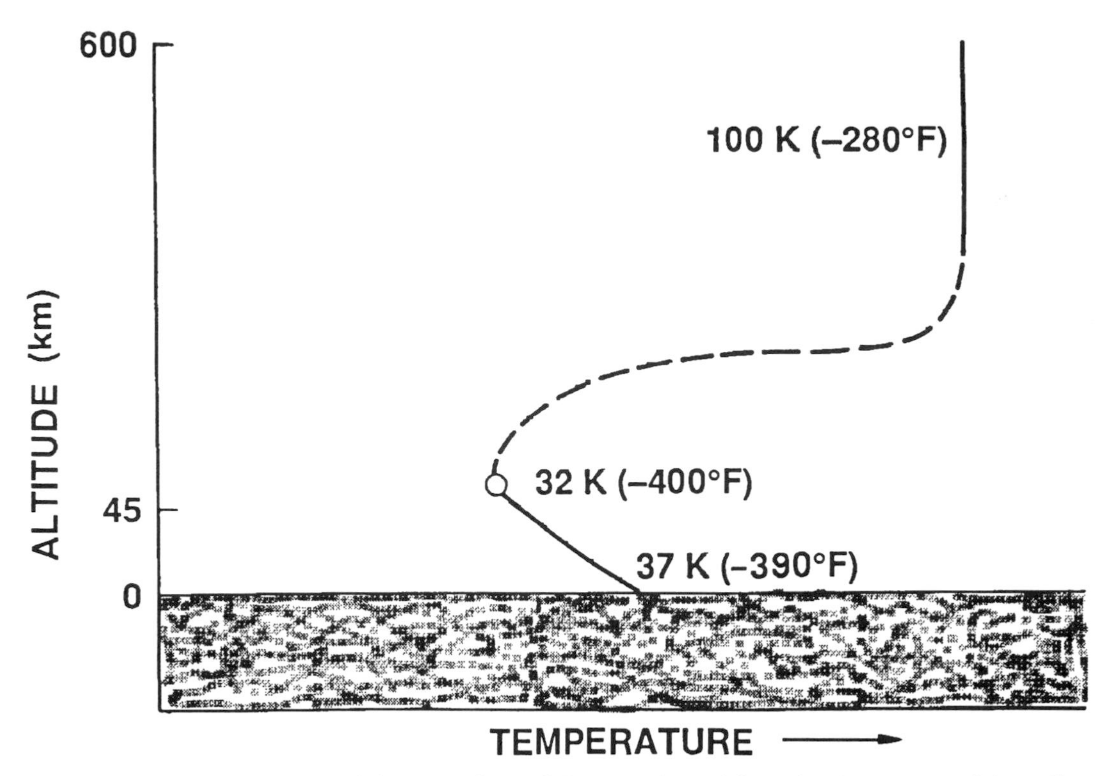

FIGURE 6.17 Temperature of the atmosphere of Triton as derived from the observations of the stellar occultation and radio occultation.

liquefied gas, most likely nitrogen, is relieved of pressure and rapidly vaporizes. The nitrogen does not have to be buried deep below the surface for this mechanism to operate when a small amount of heat allows an ice cap to be breached. Gases rush from beneath the surface in a geyser-like action carrying with them dark particles of unknown composition. Some of these particles are sufficiently fine to remain suspended in the atmosphere long enough to make the 90-mile-long, 3-mile-wide cloud.

The image on which the dark plume is shown in the figure was taken on August 24 from a distance of 62,000 miles (99,920 km). The discovery was the result of examining several images stereoscopically of the same region viewed from different angles and corrected for perspective. Later a second plume with a somewhat wider cloud was discovered east of the first, and further examination of the images leads to the conclusion that there are possibly other plumes. All are located in the central part of the polar cap region.

Because the plume material rises vertically to the 5-mile (8-km) height before spreading along as a cloud, the atmosphere is assumed to have some kind of stratification, such as an inversion layer, at that altitude. A wind also appears to arise at this altitude in contrast to what appears to be a relatively calm atmosphere beneath, in which the vertical plume rises comparatively undisturbed.

If all plumes are confined to the regions receiving most solar energy, as seems likely from analysis of the images so far, their driving force may be solar rather than internal geothermal from tidal heating of the satellite's interior. A mechanism suggested by some members of the imaging team is for layers of nitrogen ice to trap solar radiation

FIGURE 6.18 *Voyager* discovered detached limb hazes in the atmosphere of Triton. The principal layer seen in this image begins at an altitude of two miles (three km) above the surface and is about two miles (three km) thick. Fainter upward extensions of the haze reached at least nine miles (14 km). The haze may consist of organic molecules produced by irradiation of methane in Triton's atmosphere. (NASA/JPL)

similar to a greenhouse effect where it is absorbed in the darker material beneath. High pressures would be generated beneath the ice, which could ultimately be released by the nitrogen gas moving beneath the ice cap to vents or by explosively rupturing the nitrogen ice cap itself. In either case the escaping nitrogen gas could carry particles of dark material and ice in a geyser-like plume whose high temperature would result in convection carrying the particles high into the atmosphere. The activity on Triton was quite unexpected for such a distant and cold world. The outer Solar System continues to provide us with surprises and it seems a great pity that we have no further missions planned for exploring these most distant planets Uranus, Neptune, and Pluto.

Today Triton is extremely sparsely cratered compared with the highlands of Earth's Moon, implying that those few impact craters now seen on the satellite's surface have been produced during the last billion years or so. Crater densities on the most heavily cratered regions of Triton at low latitudes in the leading hemisphere are comparable with crater densities on the lunar maria. By contrast the heavily cratered surfaces of the small satellites, 1989N1 and 1989N2, appear to have been cratered in the period of heavy bombardment some 3.5 billion years ago. Unfortunately the images of Nereid do not provide sufficient detail to see craters on that satellite. Its unusual orbit suggests, however, that, like Triton, it may have been captured by Neptune rather than being a member of the original Neptunian system. Triton also would have suffered impacts in that same era, but the evidence of these is now almost completely obliterated by subsequent tectonic activities and the covering of any ancient surface by volcanic flows.

Similar to the other giant outer planets, Neptune has captured many comets. It seems most likely that comets have played an important part in the impact history of Triton, probably accounting for most of the impact craters during the past few billion years. Possibly at least a million cometary nuclei with diameters of more than 1 mile are today captured by Neptune; that is, they have the aphelia of their orbits close to the orbit of Neptune. Many of these comets must collide with Neptune and some with its satellites so that satellites are occasionally broken into smaller bodies. The smaller

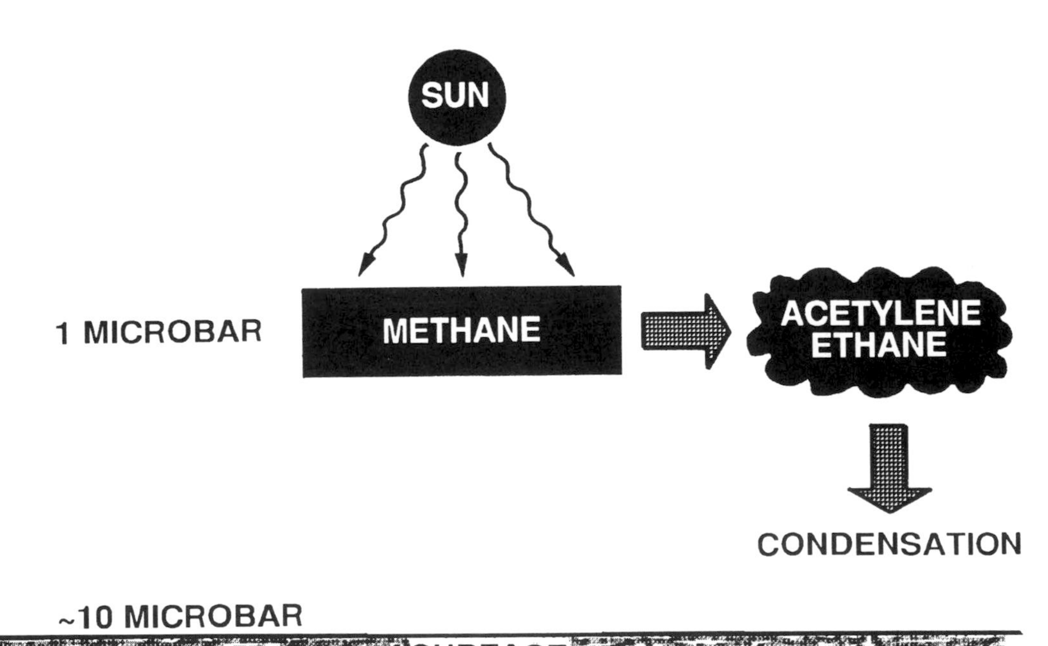

FIGURE 6.19 Possible effects of sunlight on methane in Triton's atmosphere to produce more complex hydrocarbons which may cause haze or be deposited by condensation to darken parts of the surface of the satellite.

FIGURE 6.20 A parting shot of Triton at 56,000 miles (90,000 km) beyond the satellite. Only a thin crescent of Triton's south polar region is visible to *Voyager*'s cameras. (NASA/JPL)

Neptunian satellites of today are thought to be the remnants of a larger body that was disrupted billions of years ago possibly by the impact of a planetesimal slightly larger than the one that produced the 90-mile diameter impact crater on 1989N1.

An important event during the Triton encounter was an occultation by the satellite of the star Beta Canis Majoris as observed from the spacecraft (figure 6.16, p. 130). The light from the star was recorded by the ultraviolet spectrometer and the photopolarimeter instruments as it passed through deeper and deeper regions of Triton's atmosphere both at entry and egress. The surface of Triton was arbitrarily defined as the point at which the star's light was cut off and reappeared. At this point the measurements determined that the surface temperature was 37 K (−454° F). At higher altitudes, after a slight dip, the temperature increased and reached 100 K (−280° F) at 370 miles (600 km), (see figure 6.17, p. 131). Some heating of the upper atmosphere must be taking place; possibly by absorption of solar energy or by interaction with the magnetosphere, or both.

Methane is abundant in the satellite's atmosphere at 30 miles (45 km) altitude, but the major constituent is nitrogen. Nitrogen at the temperature on Triton is subject to vapor pressure equilibrium. Associated with nitrogen ice or liquid there are nitrogen molecules in the gas phase whose abundance depends upon the temperature. If heated, nitrogen ice evaporates and gas pressure is raised. If the temperature falls, gas condenses and the pressure is reduced. For a given temperature experimenters can predict what the pressures of nitrogen and methane will be.

The atmosphere of Triton is exceedingly thin compared with those of other planetary bodies; one thousandth the pressure of Mars' atmosphere which, in turn, is somewhat less than one-hundredth that of Earth's. The imaging team made major discoveries about the atmosphere of Triton, finding that in spite of the low atmospheric pressure of 0.01 millibars, there are clouds, hazes and winds. The haze (figure 6.18, p. 132) extends all around the satellite as shown in images taken at the crescent phase after the closest encounter. It reaches to 18 miles (30 km) above the surface, and is most probably made up of smog-like particles produced by photochemical action on the methane of the atmosphere (figure 6.19, 132).

Clouds were seen over the south polar cap, which is now subliming as the region is presented to the Sun. Some clouds were detached from the surface, others extended to the surface. Clouds were seen extending across the terminator on an image of the crescent-phased Triton as the spacecraft sped away from the Neptunian system (figure 6.20). Although the temperature at the surface of Triton is 37 K (−393° F) on the south polar cap, the temperature of the exosphere reaches 100 K (−280° F) at about 370 miles (600 km). The surface temperature was confirmed within a degree by several different experiments. Triton has the coldest surface of any planetary body in the Solar System, only 37 degrees above absolute zero.

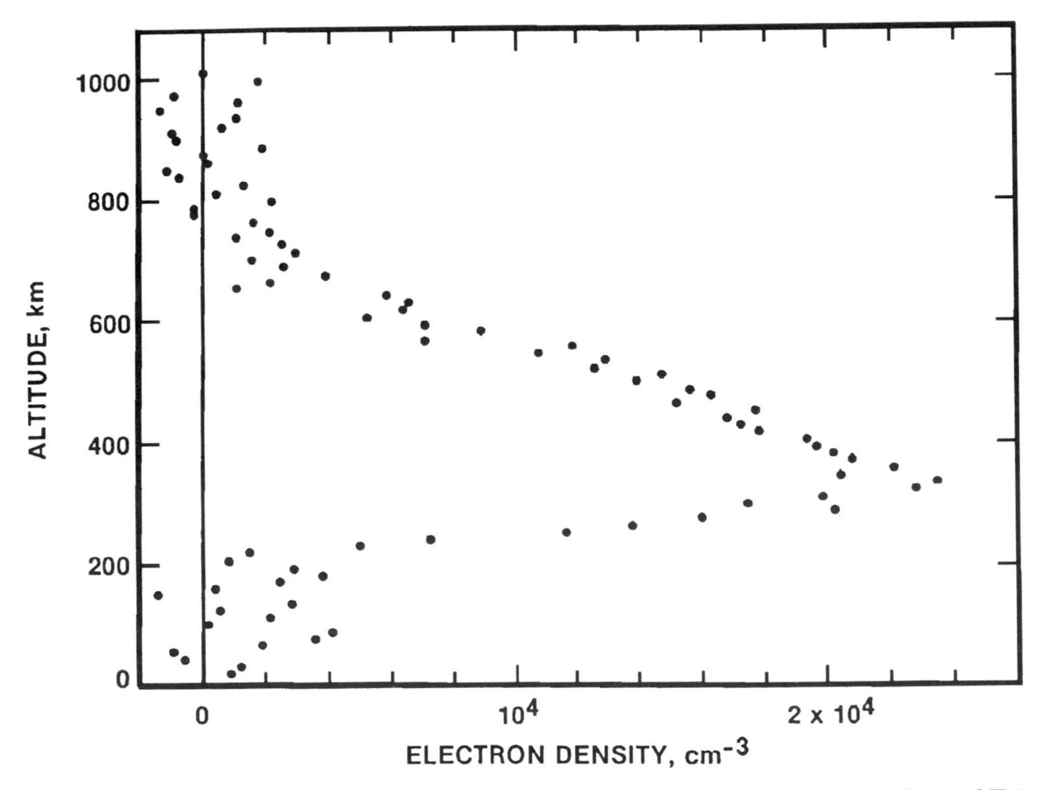

FIGURE 6.21 The radio occultation experiment provided information about the ionosphere of Triton with the electron density peaking at an altitude of just under 250 miles (400 km), as shown in this plot of the data. (JPL)

FIGURE 6.22 This dramatic view of the crescents of Neptune and Triton was acquired by *Voyager* about three days after the closest approach to Neptune when the spacecraft was plunging southward at an angle of 48 degrees to the plane of the ecliptic and heading toward interstellar space. (NASA/JPL)

TABLE 6.2 Triton and Pluto Compared

	Triton	*Pluto*
Present Distance from Sun, (AU)	30	29
Rotation period, Earth days	5.9	6.4
direction	retrograde	retrograde
Diameter, miles	1680	1450
kilometers	2705	2330
Density, grams per cubic centimeter	2.07	2.1
Albedo, percent	70	40
Temperature, kelvins	37	50
Atmospheric pressure, millibars	0.01	0.01
Surface Composition	N_2 ice, CH_4 ice	CH_4 ice +?
Atmospheric Composition	N_2, CH_4	CH_4 +?

The radio occultation experiment also provided data about the ionosphere of Triton. The density of the atmosphere was measured and the electron density determined from an altitude of 620 miles (1000 km) down to the surface. A peak electron density occurred at 217 miles (350 km), and the ionosphere appeared to end about 125 miles (200 km) above the surface (figure 6.21, p. 135). Heat coming into the atmosphere from solar radiation is absorbed mostly at this region of maximum electron density.

The temperature of the surface and the state of the atmosphere indicate that nitrogen does not pass through the liquid state as it is heated. It sublimes directly into the atmosphere. Accordingly, planetologists now believe it most unlikely that there are lakes of liquid nitrogen on Triton as had once been suggested. During the encounter, as the images showed large flat areas surrounded by scarps (see figures 6.7 and 6.8), there were speculations that these were frozen lakes of nitrogen. However, increased resolution showed details on the surfaces of these flat areas which suggest that they are more likely basins that have been flooded by material with the consistency of fluid lavas, similar to the lava filling of terrestrial calderas but, of course, by ice flows instead of rock flows.

The ultraviolet spectrometer (UVS) data recorded the presence of auroras on Triton believed to be produced by particles from the radiation belts of Neptune through which Triton passes. The emissions from Triton's auroras are similar to those observed at Titan, Saturn's large satellite.

The flyby of Triton was a little too far away from the satellite to determine if it possesses an intrinsic magnetic field, but Triton's passage through the field of Neptune results in disturbances to Neptune's magnetosphere. As a result of the flyby, the gravity field and density and other physical characteristics of Triton could be compared with Pluto and other large satellites of the outer planets. The density of Triton is greater than that of Ganymede, Callisto, and Titan and close to that of Pluto. Its value is 2.07 grams per cubic centimeter; that is, about twice that of water. The mass of Triton is calculated to be about 5×10^{-3} that of Neptune.

A comparison of the physical characteristics of Triton and Pluto given in table 6.2 shows a striking similarity between the two worlds. Pluto is warmer than Triton not just because it is now closer to the Sun but also because it has a much lower albedo, so that a larger proportion of incident solar radiation is absorbed by Pluto than by Triton. Triton is, indeed, brilliantly light-colored with some parts reflecting almost 100 percent of the incident sunlight. Even the parts appearing dark on the images are brighter than the surface of Pluto. Why the surfaces of the two bodies, which are thought to have had some common origin in another part of the Solar System away from where Neptune was formed, are so different is unknown. Carbon in the outer Solar System nebula is thought to have been present as carbon monoxide in preference to methane. If Pluto also has a predominantly nitrogen atmosphere like that of Triton, another major question is why there should be so much nitrogen rather than the expected more prevalent carbon monoxide of the outer regions of the Solar System. Another mystery is that carbon monoxide has not been detected on Triton or on Pluto, and while methane is present on Triton and Pluto none has been detected so far on Charon. Observations with the Hubble Space Telescope may resolve some of these questions in the coming years.

A parting shot of Neptune and Triton by *Voyager* (figure 6.22, p. 136) was taken when the spacecraft was approximately 3 million miles (5 million km) beyond the Neptunian system. This unique image of Neptune and Triton shows both bodies as fine crescents; the view from a spacecraft heading out of the Solar System and reaching for the beckoning stars. Undoubtedly one of mankind's greatest voyages of exploration into the unknown.

BIBLIOGRAPHY

BOOKS

Bergstrahl, Jay T., ed. *Uranus and Neptune.* Washington, D.C.: NASA Conference Publication 2330, 1984.

Burgess, Eric, and Richard O. Fimmel, William Swindell. *Pioneer Odyssey. Encounter With a Giant.* Washington, D.C.: NASA SP-349, Government Printing Office, 1975.

Burgess, Eric, and Richard O. Fimmel, James Van Allen. *Pioneer Odyssey: First to Jupiter, Saturn, and Beyond.* Washington, D.C.: NASA SP-446, Government Printing Office, 1980.

Burgess, Eric. *By Jupiter.* New York: Columbia University Press, 1982.

Burgess, Eric. *Uranus and Neptune. The Distant Giants.* New York: Columbia University Press, 1988.

Cooper, Henry S. F. Jr. *Imaging Saturn.* New York: Holt, Rinehart and Winston, 1982.

Hunt, Gary, ed. *Uranus and the Outer Planets.* Cambridge: Cambridge University Press, 1981.

Hunt, Gary, and Patrick Moore. *Atlas of Uranus.* Cambridge: Cambridge University Press, 1989.

Kohlhase, Charles. *The Voyager Neptune Travel Guide.* Pasadena: Jet Propulsion Laboratory Publication 89-24, 1989.

Morrison, David C., and Jane Samz. *Voyage to Jupiter.* Washington, D.C.: NASA SP-439, Government Printing Office, 1980.

Morrison, David C. *Voyagers to Saturn.* Washington, D.C.: NASA SP-451, Government Printing Office, 1982.

Nourse, Alan E. *The Giant Planets.* New York: Franklin Watts, 1974.

Palluconi, Frank Don, and Gordon H. Pettengill, ed. *The Rings of Saturn.* Washington, D.C.: NASA SP-343, Government Printing Office, 1974.
Washburn, Mark. *Distant Encounters.* Washington, D.C.: Harcourt, Brace, Jovanovich, 1983.

REPORTS IN SCIENCE, THE MAGAZINE OF THE AMERICAN ASSOCIATION FOR THE ADVANCEMENT OF SCIENCE

Pioneer 10. Vol. 183, pp. 301–324, January 25, 1974.
Pioneer 10 and Pioneer 11. Vol. 188, pp. 445–477, May 2, 1975.
Voyager 1 at Jupiter. Vol. 204, pp. 913–924; 945–1008, June 1, 1979.
Voyager 2 at Jupiter. Vol. 206, pp. 925–996, November 23, 1979.
Pioneer Saturn. Vol. 207, pp. 400–453, January 25, 1980.
Voyager 1 at Saturn. Vol. 212, pp. 159–243, April 10, 1981.
Voyager 2 at Saturn. Vol. 215, pp. 459; 499–594, January 29, 1982.
Voyager 2 at Uranus. Vol. 233, pp. 1–132, July 4, 1986.
Voyager 2 at Neptune. Vol. 246, pp. 1369; 1417–1501, December 15, 1989.

GLOSSARY

ACCRETION: collection of small bodies into larger bodies by mutual gravitational attraction.
ACCRETION LIMIT: radial distance from a planet inside of which accretion cannot take place.
ALBEDO: a measure of the ability of the surface of a planet or satellite to reflect radiation incident upon it. In visual light an albedo of 1 is a pure white and an albedo of 0 is an absolute black.
ALPHA PARTICLE: nucleus of a helium atom.
ANGSTROM: a measure of wavelength; 10^{-8} cm.
APHELION: most distant point in the orbit of a body orbiting the Sun.
AURORA: upper atmosphere glow produced by charged particles precipitated along magnetic field lines.

BAR: a unit of pressure; Earth's mean atmospheric pressure at sea level is 1.01325 bar.
BIT RATE: the rate at which digital information—ones and zeros—is transmitted over a communications link.
BOW SHOCK: a discontinuity where the solar wind is abruptly slowed by the magnetic field of a planet.

CALDERA: a large roughly circular flat-floored depression surrounded by cliffs formed by volcanic activity followed by a sinking of a lava lake as the eruption ended.
CARBONACEOUS MATERIAL: material rich in carbon.

CASSEGRAIN TELESCOPE: a type of telescope in which a secondary mirror reflects the incoming light through a hole in the center of the primary mirror.
CELESTIAL EQUATOR: the projection of the plane of Earth's equator on the celestial sphere.
CHARGE COUPLED AMPLIFIER: an advanced electronic detector which amplifies very faint light to produce useful images.
CUSP: the region of a magnetic field above the magnetic poles where magnetic field lines pass down to the surface of the planet.

DIFFERENTIATION: the gravitational separation of the material of a planet into shells with the heavier materials sinking toward the center of the planet.
DOPPLER EFFECT: the shift in frequency of a signal caused by relative motion between the transmitter and the receiver.
DYNAMO CURRENT: a flow of electrical current within a planet which according to the accepted theories of planetary magnetism generates the magnetic field of the planet analogous to a dynamo.

ECLIPTIC: the projection of the plane of Earth's orbit on the celestial sphere.
ECLIPTIC PLANE the plane of Earth's orbit around the Sun.
ELECTRON VOLT: energy acquired by an electron in accelerating through a potential difference of one volt.

GAUSS: a unit of magnetic induction amounting to one dyne per unit magnetic pole.
GRATING SPECTROMETER: a spectrometer which uses a surface which is etched with many fine parallel lines to produce a spectrum of radiation.
GREENHOUSE EFFECT: trapping of solar radiation within an atmosphere by gases which permit radiation to enter the atmosphere and raise its temperature but prevent the escape of heat back into space.

HADLEY CELL: rotation of atmospheric gases from equator to pole at altitude and then back to the equator at a lower altitude.
HELIOPAUSE: the boundary between the interstellar plasma and the solar wind.

IMAGE MOTION COMPENSATION: a technique to move a camera system to compensate for the movement of the object being imaged during the exposure.
IMPACT BASIN: a large circular depression surrounded by concentric mountain ranges caused by the impact of a body of relatively large size into a planet or a satellite. Mare Imbrium is an example of a large impact basin on our Moon.
ION: an atom or molecule possessing an electric charge by addition or loss of one or more electrons from its normal uncharged state.
IONOSPHERE: region of ionized gases in the low density upper atmosphere of a planet resulting from incoming solar radiation producing ions and free electrons from the atmospheric gases.

MAGNETOPAUSE: the boundary between the shocked solar wind and the magnetic field of the planet.
MAGNETOSPHERE: the region around a planet where the magnetic field of the planet is dominant and holds off the solar wind.
MAGNETOTAIL: the region of the magnetosphere extending from the planet like a windsock trailing down the solar wind.
MAGNITUDE, STELLAR: the apparent brightness of a celestial object. Each magnitude is a

change of 2.5 times in brightness. The faintest star visible to the unaided eye is magnitude 6. Brightest stars are magnitude 1, with a few even brighter. Planets such as Venus attain magnitudes of less than one and such magnitudes are expressed in negative values.
MASER: an amplifier based on the use of microwave amplification by stimulated emission of radiation. Energy stored in a molecular system is stimulated by the input signal.
MILLIBAR: a unit of pressure equivalent to 1000 dynes per square centimeter. It is approximately 1/1000th of Earth's surface atmospheric pressure.

OCCULTATION: the obscuration of a celestial body by one of larger size, such as when a star appears to pass behind the Moon.

PERIHELION: the point in a planet's orbit which is nearest to the Sun.
PERTURBATIONS: disturbances to the regular motion of a celestial body caused by the presence of other bodies.
PHOTOCHEMISTRY: chemical reactions caused by the action of radiation such as that from the Sun.
PIXEL: picture element; a small square or rectangular area with its own shade of grey, many of which make up the complete image.
PLANETESIMAL: a body made up of condensates from a primordial nebula from which planets and their satellites formed by accretion.
PLASMA: an electrically conductive gas consisting of neutral particles, ions, and electrons which as a whole is electrically neutral.
POLARIZATION: the state of radiation when transverse vibrations occur in a regular manner such as in a plane or in a circle. Polarization can occur at the source or as a result of the radiation passing through materials or being reflected by materials.
PRECESSION: change in the axis of rotation of a spinning body such as a planet, caused by a torque produced by the gravity of other bodies.
PRINCIPAL INVESTIGATOR: the team leader of a group of scientists associated with a particular experiment.

RADIATION BELT: an envelope of charged particles trapped in the magnetic field surrounding a planet. The first such belt, one surrounding Earth, was discovered by James Van Allen.
RETROGRADE: motion that is opposite to the usual direction of celestial bodies in a system.
ROCHE LIMIT: the radial distance from a planet at which a satellite could be pulled apart by gravitational forces.

S-BAND: a radio frequency band allocated to space communications. *Voyager* used a frequency of 2.3 GHz within this band. *Pioneer* used 2.1 and 2.3 GHz.
SHEAR ZONE: the layer between two masses of atmosphere moving relative to each other.
SHEPHERDING SATELLITE: a satellite in orbit close to a ring which prevents ring particles from straying radially in or out from the ring.
SOLAR WIND: stream of plasma flowing outward from the Sun.
SUBLIMATION: change in state from a solid to a gas without passing through a liquid state.
SUBSOLAR POINT: location on a planetary body at which the Sun is overhead.
SYNCHROTRON RADIATION: radiation developed from electrons oscillating within a magnetic field.

TECTONICS: molding of a planetary surface by forces acting from within the planet.
TELEMETRY: making measurements at a distant location and transmitting the data to another location for analysis.

TERMINATOR: the boundary between the sunlit and dark hemispheres of a planet or a satellite.
TRANSIT: the apparent passage of a celestial body across the face of another body without eclipsing that body.
TROJAN COMPANION: a body moving in the same orbit as a satellite in a stable configuration at a Lagrangian equilibrium point. Named after a group of asteroids orbiting ahead and behind Jupiter at such points, asteroids which were named after the heroes of the Trojan War.

VIDICON: a television imaging device that uses a photoconductor as a sensor.

X-BAND: a radio frquency band allocated to space communications. *Voyager* used a frequency of 8.5 GHz within this band.

ZODIACAL CONSTELLATION: one of twelve constellations along the projection of the ecliptic plane on the celestial sphere and within which the planets appear to move as observed from Earth.

INDEX

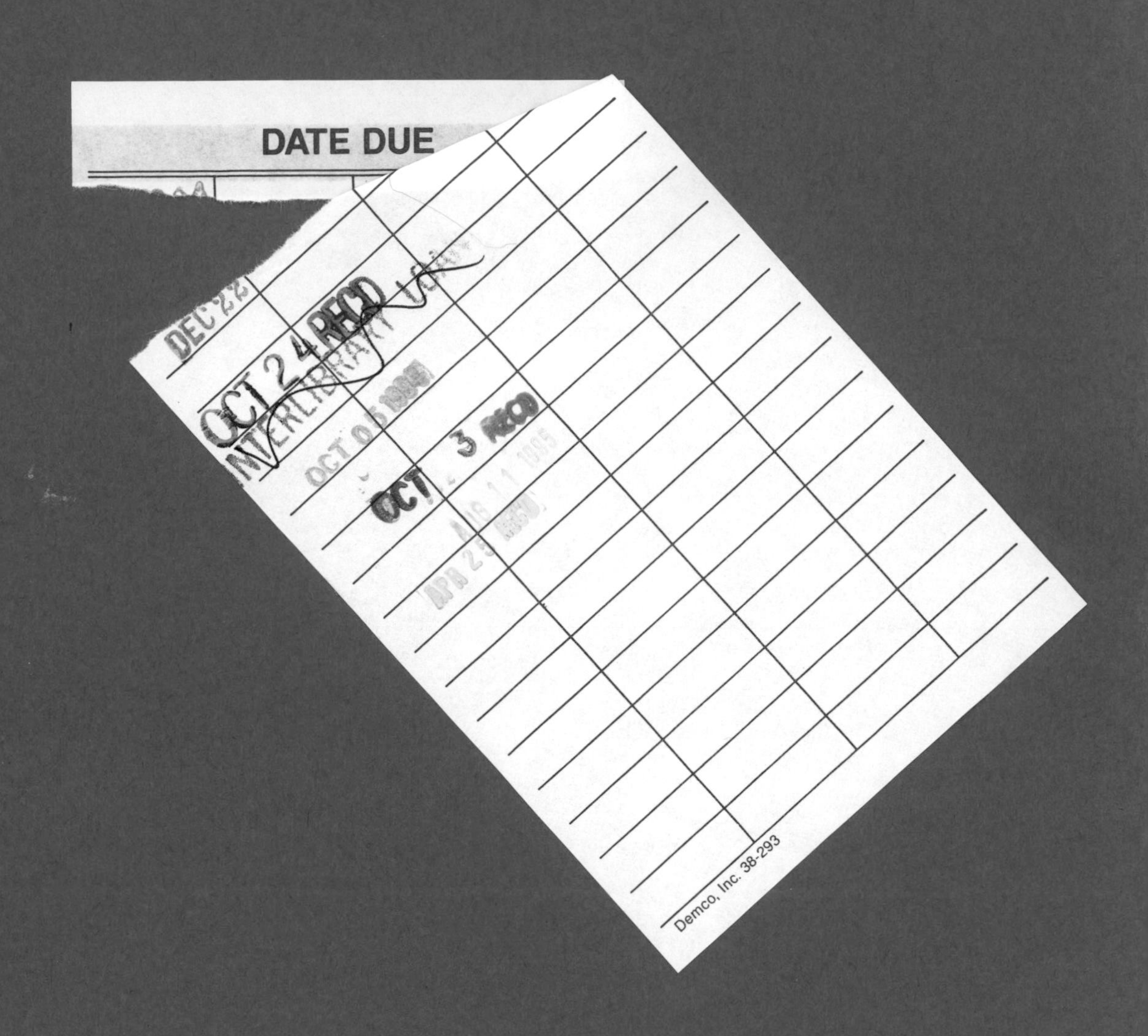

DATE DUE
Demco, Inc. 38-293